Fragmented Society: The Diffusion of ICT and China's Modernization

Hefan Xu

# Fragmented Society: The Diffusion of ICT and China's Modernization

**Bibliographic Information published by the Deutsche Nationalbibliothek**
The Deutsche Nationalbibliothek lists this publication in the Deutsche Nationalbibliografie; detailed bibliographic data is available online at http://dnb.d-nb.de.

**Library of Congress Cataloging-in-Publication Data**
A CIP catalog record for this book has been applied for at the Library of Congress.

Zugl.: Heidelberg, Univ., Diss., 2017

ISBN 978-3-631-77176-1 (Print)
E-ISBN 978-3-631-77241-6 (E-PDF)
E-ISBN 978-3-631-77242-3 (EPUB)
E-ISBN 978-3-631-77243-0 (MOBI)
DOI 10.3726/b14938

© Peter Lang GmbH
Internationaler Verlag der Wissenschaften
Berlin 2018
All rights reserved.

Peter Lang – Berlin · Bern · Bruxelles · New York ·
Oxford · Warszawa · Wien

All parts of this publication are protected by copyright. Any utilisation outside the strict limits of the copyright law, without the permission of the publisher, is forbidden and liable to prosecution. This applies in particular to reproductions, translations, microfilming, and storage and processing in electronic retrieval systems.

This publication has been peer reviewed.

www.peterlang.com

# Contents

**Abbreviations** .................................................................................. 9

**Introduction** .................................................................................. 11

**1 The Fragmented Society of China** ........................................... 15
    A Train Metaphor ........................................................................ 15
    The Carriages of the Train—*hukou* ........................................... 17
    Urbanization ................................................................................ 19
    Passengers who are Influenced by the One-Child policy ......... 22
    The Environment of the Train .................................................... 25
    Conclusion ................................................................................... 28

**2 The Eastern Media Philosophy** ................................................ 31
    The Role of Intellectuals in Traditional Chinese Culture ......... 32
    The Intellectual, Liberty and Freedom in China from 1880 to 1949 ......... 35
    The Introduction of Liberty and Freedom to China ................. 39
    New Concepts without Real Meaning ....................................... 42
    Conclusion ................................................................................... 46

**3 The Two Stages of Development of the Media in China after 1949** ........................................................................... 47
    The Outside Circle: Interaction of the Media and Society ....... 48
    The Inside Circle: Social Condition and the Media .................. 50
    Mao's Period—Slogans and Social Mobilization ....................... 51
    The Post-Mao Period: Learning Experience and Fast Diffusion of ICT .... 60
    The Learning Experience and Telecommunication Reforms ... 69

## Contents

    The Fast Diffusion of ICT .................................................................. 78

    Conclusion ........................................................................................... 94

**4 Is the Diffusion of ICT in China Special?** .................................... 97

    The Diffusion Curve of ICT in China ............................................... 97

    A Comparative Analysis of ICT ........................................................ 100

    Conclusion ........................................................................................... 113

**5 Losing and Rebuilding the Social Trust System** ........................ 115

    The SARS Virus .................................................................................. 116

    Guo Meimei ........................................................................................ 120

    Conclusion ........................................................................................... 125

**6 How an Indigenous Society Develops in the Wave of Modernization** .............................................................................. 127

    A Window to Observe ....................................................................... 127

    The Difficulty of Entering into the Local Community .................. 128

    A Traditional Society ......................................................................... 132

    The Pervasive Usage of ICT in the Minority Area ......................... 137

    The Tourism Modernization Process .............................................. 142

    Live or Achieve—Who is the Owner of Modern Society? ........... 161

    A Dialog with a Local Intellectual ................................................... 162

    Conclusion ........................................................................................... 166

**7 Reflections on Indigenous Culture and Modernization** ........... 169

    Behind the Fast Speed of Economic Growth ................................. 170

    The Digital Divide: The Usage Gap and the Age Divide .............. 171

    Different Carrier— Tradition and Modernity ................................ 174

Cultural Awareness ................................................................................ 176
Conclusion ............................................................................................ 181

## 8 Summary ........................................................................................ 183
The "Social Vacuum System" with Fragmented Social Memory ................ 184

## List of Figures ..................................................................................... 193

## List of Tables ...................................................................................... 195

## Bibliography ...................................................................................... 197

# Abbreviations

| | |
|---|---|
| B2B | Business-to-Business |
| CASS | Chinese Academy of Social Sciences |
| CDMA | Code Division Multiple Access |
| CERNET | Computer Emergency Response Team |
| CNNIC | China Internet Network Information Center |
| DLP | Digital Light Processing |
| ESDQMDAP | Economic and Social Development of Qiandongnan Miao and Dong Autonomous Prefecture |
| ESnet | Energy Sciences Network |
| GDP | Gross Domestic Product |
| GPRS | General Packet Radio Service |
| ICA | Institute of Computer Application |
| ICT | Information Communication Technology |
| ITU | International Telecommunication Union |
| IWS | Internet World Stats |
| LCD | Liquid Crystal Display |
| MIIT | Ministry of Industry and Information Technology of China |
| NBSC | National Bureau of Statistics of China |
| NCFC | National Computing and Networking Facility |
| NSF | National Science Foundation of US |
| NTIA | National Telecommunications Infrastructure Administration |
| OEM | Original Equipment Manufacturer |
| PDP | Plasma Display Panels |

# Introduction

In 2010, I participated in a project to investigate how modern media is changing traditional Chinese villages. This first look at indigenous Chinese culture piqued my curiosity about the relationship between media and modernization. I was shocked when I saw how people carry heavy loads along the narrow mountain road, and at first wondered why they do not choose a flat place to build apartments. I still wonder how modernization processes are changing the area, and what the relation is between tradition and modernity. To this day, I still remember my first impressions of that village: the traditional wooden buildings (*diaojiao lou*), handmade clothes, majestic terraced mountains, friendly local people, and the ubiquity of digital devices. This is an ideal place to observe the media and modernization of China, and serve as a window to understand social change.

China, as the largest developing country, the biggest socialist country, the fastest rising GDP country, the second largest country by land area, the most populous country, the country with the second largest economy and a single party country etc., has experienced rapid transition. In recent years, increasing attention has focused on China, as it is regarded as an economic engine for the world and the center of digital device production.

So far, information and communication technology (ICT) has boosted economic development in China. The process of modernization can be seen in the widespread use of energy-saving tools, and with physical devices that perform more efficiently and effectively than humans (Levy, 1966).

At the moment, in China as well as in other countries, traditional media are being replaced by new media, especially the ICT that combines computer, broadcasting, satellite, and visual technologies; for example, computers and smartphones have merged books, news, music, television, radio, etc. into one device. In the information age, individuals are adapting to new technologies as well as using them.

As technology diffuses across the globe, developing countries are considered to be "hitching a ride" on innovation, acquiring advanced techniques while investing less in new technology. The ability of so-called "catch-up countries" to benefit from mature technologies to accelerate the penetration procedure is called "the advantages of backwardness" (Glaziev, 1991; Keller, 2001). As a catch-up country, China is trying to modernize quickly, and will have to face many issues.

"A society will be considered more or less modernized to the extent that its members use inanimate sources of power and/or use tools to multiply the effects of their efforts" (Levy, 1966, p. 11). To understand modernization and social change, it is important to find out what triggers them. To understand China's modernization, it is necessary to take into account the development of ICT.

Toffler, Longul and Forbes (1981) first introduced the concept of the three revolutions, namely the agricultural revolution, the industrial revolution and the information revolution. As ICT has become more widespread, society has entered the information era, which means the diffusion of ICT has set off a new wave of modernization.

However, this wave did not appear suddenly; it can be traced back to the beginning of the industrial revolution. *In A History of Mass Communication: Six Information Revolutions*, I. E. Fang (1997) attempts to divide the Information Revolution into six stages according to their characters, namely the Writing Revolution (eighth century B.C.), the Printing Revolution (the second half of the fifteenth century), the Mass Media Revolution (middle of the nineteenth century), the Entertainment Revolution (the end of nineteenth century), the Communication Toolshed Home (the middle of the twentieth century, when the home became the central location for receiving information and entertainment), and the Information Highway. As can be seen from the dates given, each stage is shorter than the one that preceded it, with the last four revolutions occurring in the past two centuries. In an information society, changes in industry and human life are closely related to the Internet, where information appears and disappears rapidly, and new technologies develop ever faster.

The six information revolutions that Irving Fang highlighted comprise a wide spectrum of information and media. The initial period of each revolution suggests how long it takes until the older revolution transfers into the new revolution. The information revolution accelerates to transform into the next stage. It took approximately 1,700 years for the first revolution to transition into the second revolution. However, the last four information revolutions, impressively, have finished their transition and overlapped during the past two centuries. In the information society, the update of the industry and lives of human beings are closely related to the Internet. It will concurrently create and delete information at a high speed. Most importantly, the update of new technologies will develop rapidly in the information society.

Modernization is a process that transforms a traditional or pre-technological society into a society characterized by machine technology, rational and secular attitudes, and highly differentiated social structures (Huntington, 1971; Tipps,

1973). The diffusion of ICT is not identical with "modernization", nor is it the main cause of China's modernization. However, in view of how widespread ICT has become in China, and the information age is labeled as the new revolution of the world, it is important to understand the social transformation process in one of the oldest cultures and longest-flourishing civilizations in the world.

# 1 The Fragmented Society of China

## A Train Metaphor

China, as the largest developing country, is transforming from tradition to modernity, and is now experiencing conflictual social change. During this transformation, the development of ICT is influenced by economic development, social structure, and the political system; ICT diffusion also has a reciprocal effect.

A general impression about the China train is that it resembles a steam locomotive train. As the train is gathering speed, it rumbles down the tracks, with increasing noise, smoke, dust and dirt. Green things near the track are covered with soot; it is not an elegant or clean system. Nevertheless, the train is accomplishing its main function—transportation. The train is not the most efficient but it runs with lower cost.

If China's society is like a train, several factors need to be considered about this train, namely time, space, people, and communication. The train becomes more crowded over time: think of the passengers as members of society, and each destination is a step toward an improved living standard and the next stage of the planned economy (arrival times can be attributed to the stages, e.g., 5-year plan, 10-year plan, etc.). The passengers and the organ of the train come to the consensus that the train is uncomfortable, but are pleased that it is doing the job, getting them to the next destination.

At the same time, during a trip, especially an uncomfortable trip, the role of entertainment tends to be highlighted and media become important tools for recreation and passing time, which means, the media at this early steam locomotive stage not only serve as a commodity but also play other social functions, such as maintaining social balance.

The train decreases the travel time between departure and terminus and media play a role serving the people in the train. If regarding the Chinese society as a fast – moving train, to some extent, the users can be distinguished into two dimensions based on whether they use media or not. Passengers who use the media have possibilities to know what happened outside of the train.

The intervention of media makes passengers communicate less with each other, care less about the current circumstance, but focus more on the digital devices which are held in their hand. Passengers in the same train might have the opportunity to discuss the conditions or the circumstances of the train or other topics of interest, but because the majority of people are more interested in the media than the train surroundings, there is little direct personal communication. However, there are also fewer chances of face-to-face conflict, increasing the overall stability of the train. Therefore, the diffusion of ICT divides the society into two parts, people who use it to get information, and people who do not.

Among the media users, people who use different media tend to get different information. More surprising, the difference even exists within the same website, the content displayed depends on which access people use, for example, a website viewed on a computer will have different content than a mobile phone terminal. Therefore, people can be divided into several categories according to the media they use.

The first group is who can use the Internet and interact with the media. These users are not only information receivers but also information senders. They can search information, grasp information and use it. In this group, the users can be categorized into three groups, namely the new generation,[1] the working class, and the better-educated elderly people who like to learn new technology.

The second group is composed of workers. Most of them were born after 1965, and use ICT for their work. Compared to the first group, they accept ICT in a passive way; most of them are forced to learn ICT for their jobs and to earn money. However, they use the media more diversely compared to other groups, and to some extent, they are the group who witnessed the development of ICT from the beginning until now. They use office software to handle and process documents and play online games. At one time, the most fashionable online game was *toucai*,[2] which even aroused the attention of the central government. In addition, they read newspapers, listen to radios in their car, watch TV after returning home, and also read magazines and books; they use their computer to work and play games to kill time. They are the group that most widely uses both traditional media and new media in China.

The third group is the elderly people who do not want to be abandoned by the information society, so learn to use ICT. Most of them are open-minded with a good education. They like to accept the new technologies and use them. Compared to the other two groups, they interact less with ICT; only a

---

1 To distinguish them from the other groups, the new generation is called 80er, 90er, etc. The new generation is naturally intimately friendly with ICT. They are assertive, independent, like to pursue new technologies and play games.
2 An online game designed by Tencent company. A player can have their own garden and plant vegetables, they can buy animals, seeds, fertilizers and have to take care of their garden because other players will steal their vegetables. Many civil servants like to play this online game, and it is now restricted.

small percentage of them can type. The majority of the third group are passive users: surfing web pages and video with their children.

The Chinese language system is different from the Western language; however, the invention of the computer is based on the English type system. The main type styles of computer are *pinyin*[3] and *wubi*. The *pinyin* system uses the Western alphabet, which can be transformed into Chinese characters as they are typed on the keyboard based on their sound, for instance to type China (*zhongguo* 中国) on the keyboard, users should type *zhong* and *guo* Therefore, users need to know the pronunciation of each character in Western letters, otherwise, they can just read Chinese characters but cannot type. In terms of *wubi*, users need to remember the etymon of the characters. However, both of these types are difficult for the elderly to learn, even those considered well-educated. If they do not learn *pinyin* and *wubi* systematically, it is difficult for them to use the computer, and they can only browse web pages.

The people who cannot use ICT media are mostly elderly. As a result, many elderly people are kept outside the door of the information society. It is difficult for those that try to catch up to the information society since the ICT hardware and software are designed for young people. These people are like soldiers falling behind the group that has to use maximum effort to catch the new technology and adapts to the social change.

It cannot be simply concluded that the diffusion of ICT is the sole reason of social stratification, but it is certainly an accelerator. Income is an important factor contributing to social stratification, but in this information society, the skills of receiving and utilizing information are closely related to income. More and more evidence indicates that the information gap results in a new social stratification.

## The Carriages of the Train—*hukou*

As mentioned, the society of China is like a fast-moving train. The train contains many carriages. The train carriages have physically separated the train into different parts. Passengers are distributed to sit in different carriages to maintain

---

3   In 1958, the Chinese government promulgated the issue of transcribing the Mandarin pronunciations of Chinese characters into the Latin alphabet; afterwards, in 1982 the International Organization for Standardization adopted it as the international standard. In the elementary school, the first two years are mostly used to learn pinyin as it is the basic knowledge for the students to learn Chinese as well as to input Chinese characters into a computer.

an order of the train. Looking back at China's history, *hukou* is like the carriages, *hu* in Chinese means family unit, *kou* means mouth, together they mean identification. Every Chinese must register the *hukou*. The *hukou* always exists as a family unit, which means one family will have one hukou notebook. Inside the notebook is all the information of every family member. In the beginning, the *hukou* policy was aimed to restrict social mobility. However, this barrier is gradually being broken down by the urbanization and technology revolution, which changes the labor structure and increases social mobility.

During the second half of the twentieth century, concern over the population capacity of urban centers rose to consume the attention of the central government. This concern was great enough to cause the central government to decide to set up the *hukou* policy—a barrier designed to separate villages from urban areas.[4] In the 1950s, a special word, *mangliu*[5] (Li, 2008), was used to describe the labor force migrating to the cities spontaneously. More precisely, *mangliu* are workers who are not recruited by the government to work or study in the cities, which means they moved to the city illegally to find work. These workers are not protected by the law or city security system, unlike the migrant workers (*nongmin gong*, also called peasant workers) who are permitted to work. The aim of *hukou* is to prevent surplus rural laborers to work in the cities (Li & Wang, 2007; Zhang, 2013). Some of the authorized migrant workers may work decades in their new city, but still not officially belong to it. They have double identities—they are villagers but they regard themselves as city inhabitants. Their kids cannot study in cities as urban citizens' children do and they have to pay different kinds of fees for cities where they live. That is one of the important reasons that the minority people in the investigated villages, which we shall see in later chapters, return to their village if the local economy improves.

The *hukou* policy is like a visa for migrant workers who want to live and work in cities and led to serious consequences (Zheng & Lu, 2002). It not only restricted the rural villagers from entering the urban environment, but also relocated city inhabitants to the villages as well. Many well-educated youths

---

4 The *hukou* together with other policies tried to prevent immigrants from rushing into the city but did not function as the government expected. The reality is more and more work forced migration to the cities, and most of them are young and middle-aged, which I will explain in the next section.
5 In 1953, the State Council set off the policy that persuaded the villagers not to pour into the cities. The term *mangliu* came from this policy. It points to the people who migrate from villages to cities. As it includes the meaning of discrimination, recently the scholars and experts use the word *mingong* (migrant laborer) to replace *mangliu*.

who lived in the city were sent to the rural areas. As a result, China entered into an anti-urbanization period between 1958 and 1978 (Wang, 2001). As China's economy started to recover, the rate of urbanization grew more slowly than the economy. The trend of urbanization naturally correlates positively with the increase of economic performance. The segregation established by the anti-urbanization policy resulted in decreasing urbanization whilst the rate of industrialization was increasing (Meng, Zhang, & Shen, 2004; Ye & Huang, 2004).

## Urbanization

According to the world urbanization statistics, a country's urbanization is closely related to its economic performance or GDP growth. Governments have often undertaken active policies affecting the urbanization process (Bloom, Canning & Fink, 2008). The expanding urban areas of Chinese cities are much larger than other countries' not only in terms of space but also with a number of inhabitants.

The previous urbanization process was a political decision rather than stimulated by economic development. It can be categorized into three stages. The first stage of urbanization in China was from 1960–1978. Mao encouraged the youth and literate citizens to relocate to rural areas to practice, as a result, urbanization at this time fell substantially from 19.7 % to 17.9 %. The urbanization during this time was actually anti-urbanization, but as with the industrial revolution, the world experienced fast urbanization during this period, making China strikingly different. The second stage was from 1979 to 1995. As reform and opening (initiated in 1979) stimulated growth, the rate of urbanization grew gradually. Since 1980, as the government promulgated the one-child policy, the growth rate of the total population was only 0.5-1 %, but the annual urbanization growth rate was 3-5 %. It is worth noting that a substantial part of the urban population growth in China derived from rural–urban migration. The third stage is from 1996 until now where urbanization entered into a rapid but unstable period. According to the Chinese government, more than 200 million peasants now live and work in cities. At this time, China joined the WTO and conducted rural-transformation. The government abolished agricultural taxation and encouraged young people to move to the urban cores in order to transfer the surplus agricultural labor force (Chen, Liu & Tao, 2013). According to the report from the Chinese Academy of Social Sciences (CASS), by 2008 the rate of urbanization was 45 %, and China's urban population increased dramatically from 170 million in 1978 to 670 million in 2010. The rate accelerated and peaked in 2013, afterwards, the high-speed urbanization ended in 2016 (Zhang & Liu, 2010).

The period of anti-urbanization has led the current urbanization into a difficult situation. Now, the urbanization in China has to press the fast button, as the economic growth should have a compatible urbanization level. During the early and middle stage of industrialization, the market needed a sufficient labor force to promote industrial development. Meanwhile, as industrialization in China overlaps with digitalization, the intellectualized nature of technology and machinery needed less manpower. Consequently, a dilemma situation for modernizing China is that the current urbanization has to deal with providing more positions for the increasing labor force while promoting technology development needs less labor force. The modernization has to provide more jobs, however, with technological improvement, machines have replaced people, and less labor is needed in the production field.

The *hukou* policy intended to prevent rural workers from moving to the cities has proven futile, since two-thirds of China's population is now living in cities or towns and thousands of migrant laborers work in cities. The *hukou* system aims to keep rural villagers out of the urban areas, however, the truth is the modernization of China largely depends on these farm workers (Jiang, 2004). Since the end of the 1970s, when the open-up policy started, many rural laborers rushed into the cities, suggesting when the modernization began the urbanization was already started (Gu, 2004). The old urbanization strategy of establishing new villages and towns to attract the labor force struggled to realize the goal. Since then, the central government tried another routine to develop the city, that is: transfer the surplus laborers to cities by establishing super cities to hold the burgeoning population, making full use of facilities of a centralized infrastructure to optimize resources for a large number of inhabitants. This is the typical way of using physical hardware (such as establishing apartment infrastructure) to promote the development of software (such as an individual's education and their behavior).

The information society and fast urbanization brought with it a multitude of consequences. In the past, as the huge rural labor force rushed into the cities, manual labor jobs were easy to find, since China used to be at the bottom of the global industrial chain with a labor-intensive economy. China's old-fashioned, labor-intensive industries provided employment for millions of unskilled workers. However, in contemporary society, the labor-intensive industry is shrinking, and ICT and other technical skills are the basic requirements. Therefore, many people without computer skills have difficulty finding jobs.

At the beginning of 2010, the east coast of China started to recover from the economic crisis, and an unusual phenomenon called 'labor shortage' occurred. Both the central government and local governments were shocked by the new phenomenon and could not imagine that the country with the largest population was lacking an adequate labor force. However, if this phenomenon is looked at from the perspective of the information society, the 'labor shortage' reflects that the structure of the labor force is not changing at the same speed as society, which means when the society entered the information age, the society members were not equipped to face the social change.

The labor force structure has been updating and evolving since the 1990s (Yu, 2010). The aging workers who have difficulty using new technologies are being replaced by young and better-educated workers. The youth with city *hukou* was not enough to fulfill the job market as the large-scale industrialization and informationization progressed, therefore, workers without city *hukou* were permitted to work in the cities. The youth who could operate computers and with technical skills replaced the workers who lacked the abilities; as a result, less skilled city workers lost their privilege to get jobs, and because of the lack of information workers, matching job openings to qualified applicants became difficult. The transition of the society—from industrialized society to information society may invoke changes in every aspect, in terms of labor force, the shortage of labor is just an insubstantial semblance, the deep reason can be attributed to the social system not fitting the social change.

Rapid urbanization, by establishing new cities and investing in infrastructure, has made Chinese society vibrant. The process of modernization changes society immensely. Rural people can move to the urban area and benefit from the technological innovations and be protected by the city welfare system, however, this does not mean that the 'fruits' of this physical software can really serve the masses.

During the process of industrialization and urbanization, a large number of surplus rural laborers are migrating to big cities, and most villagers who are left in the rural areas are the elderly, women, children, and the disabled.[6] The popularity of television and the Internet dramatically changed the life of people left behind and these are their main access points to the information age. Compared to people in cities or developed areas, they are a vulnerable group who are weak in information acquisition. Moreover, in the western part of China, there are

---

6  In the investigated area, before the tourism economy, many villagers went to east China or big cities to work. Some of them returned when the tourism economy started. However, people who are left have different opinions about villagers who come back as they believe people who do not have the capability to live outside will return. That is, they think anyone who returned cannot adapt to the outside world and therefore is not a qualified modern person, who is then labeled as a loser.

many minorities with dialects—the local people cannot understand Mandarin even though they have physical access.[7] As the government proposed a new policy to develop the towns of China, the "hollow village"[8] is considered to be a big obstacle of urbanization.

As more people enter the urban areas and cannot benefit from the social welfare of the urban area, conflicts over social security have become acute in recent years (Chen et al., 2013). Nowadays, the elite, literate, and young are realizing ICT benefits; the illiterate and elderly are laggards, and their land is eroded by the industrial companies. After the bonus of urbanization ends, what is the future of these immigrants, what should they do in this information age since they are marginalized by the technological society?

China utilizes the low labor cost of farm workers to accumulate capital for the massive modernization process. If you look at today's China, thousands of skyscrapers are built by the farm workers whereas they are living in bad conditions and their salaries are unfair. The information society provides a chance to change this situation, since the new technologies give farm workers chances to change their living standard. Currently, many of them learn how to command the new technologies.

The large-scale movement of rural villagers to work in an urban area invokes the update of the labor force. The *hukou* results in fragmentation at the policy level, but the information society will shape a society fully fractured. However, with this change we will see another picture that people sitting in the train will lose connection with each other and have different social realities.

## Passengers who are Influenced by the One-Child Policy

In China, the young generation influenced by the one-child policy adopt the new technology almost at the same time as their Western counterparts,[9] supported by their parents and schools, they grow up in the new information society, which endows them a special capability to adjust to the information world. The wide

---

7   The investigated region is a minority area. During the fast urbanization, local adults go to urban areas for work, and their children live together with the elderly people. The children learn to use ICT very fast, and they become windows for the elderly people.
8   There are few youths and adults in the village, and the people who are left behind are elderly, children, women, and the disabled.
9   The central government attached great importance to inspiring children to learn computers.

diffusion of ICT breaks the crystalized society and gives opportunities to rural people who have the capability to operate new technologies.

People who live in rural places, villages, and towns try to give their children the best education,[10] as they believe a good education can change the children's destiny. The parents stimulate the children to learn compute so the young generations adapt rapidly to the information age—both city and village children can use computers. Meanwhile, the one-child policy is changing China into an aging society much quicker than expected The aging employees are faced with job challenges because of the new media.

As new technology spreads, especially new media such as the Internet, the traditional ways of working are changed. Instead of writing by hand, most documents are typed on computer keyboards, and machines and production lines tend to be controlled by computers as well, which means if workers cannot use the computer in the modern society, they may never find a job.

As mentioned earlier, the young generations who are influenced by the one-child policy are the main ICT users and are adept at learning new technologies. The one – child policy was introduced in 1978. After three decades, the government started to gradually loosen control of the one-child policy and now encourages families with one child to have a second child. The young generation, namely the 80ers, 90ers and 00ers have been greatly influenced by the one-child policy, and they are the driving forces of dramatic change in the information age.

The one-child policy has created a new generation that is very lonely; as a result, television, computers and digital devices have become their friends instead of siblings. The new generation is naturally intimately friendly with ICT. They are assertive, independent, like to pursue new technologies and play games. Besides, the ICT companies tend to design digital devices for the young generation, thus, the mainstream consumers are mostly the new generation. ICT for them are vectors and tools. The new generations are the most advanced group compared to other groups in terms of using ICT, as they are not only dependent on ICT for information, work, and entertainment but also exceed the passive user role and gradually advance to commanding ICT.

The young generations, which are influenced by the one-child policy, are like emperors in their families. Their grandparents and parents surround them and give them whatever they want. Some people may blame them and say that they spoil their children, and scholars argue the one-child policy has led to the

---

10   In the minority regions parents who have a better financial condition tend to send their children to study in towns or cities.

'new generations' egoism and tendency to flaunt (Feng, 2000).[11] Alternatively, this spoiled generation has benefited from tremendous opportunity compared to the previous generations, receiving more attention, resources, and cultivation. Families with only one child can spend all their resources on one high-quality education compared to large families that would have to allocate limited resources to many children.

The new generation is not only characterized by being spoiled, but also has a more independent personality compared to the last generations. Unlike the older generations, the young generations get their information mostly from the Internet.

Four decades ago, Ball-Rokeach and DeFleur (1976, p. 3) asked the question "Do mass communications have widespread effects on individuals and society or do they have relatively little influence?" To answer this question, Ball-Rokeach and DeFleur used the dependency model of mass media effects to analyze the roles media play and the relations in which mass media interact with society and individuals. By analyzing relationships among the society, mass media and individuals, they focused on three aspects, namely, the subsystem of communication, the individual's social reality and attitude formation to explore the media communication system.

They found that the mass media system is like a filter in having its own information-gathering system, processing system, and selective dissemination system (Ball-Rokeach & DeFleur, 1976; McQuail, 2010). The mass media system chooses what kind of world the masses can see and decides which culture is a popular culture and which is not.

The information transferred by the media can be taken into account when analyzing the trend of people's social reality and values. The history, customs and current situation interact together on the individual's faith, beliefs, and social realities. After having studied the essential roles of the mass media in the process of modernization, scholars believe that the media play a role to build an individual's social reality (Adoni & Mane, 1984; Glauser, 2015; Lewis & Weaver, 2016; Watson & Hill, 2015). In addition, the effects of messages from mass media flow back to influence the social characteristics (Kassarjian, 1965; Katz et al., 1973; Lundby, 2009).

According to the independent model of the mass media, the information content is variable based on the type of access. Hence, the social reality of the

---

11 Feng Xiaotian was one of the important scholars who focused on the one-child policy, completing a large-scale study on children of that period.

young generations might differ from the previous generations and their social reality will flow back to shape the social characteristics and influence the social memories.

## The Environment of the Train

Using the train metaphor to describe the society of China, after introducing the train, the carriages and the passengers, what about the environment of the train? Sun (2004) claims that the society of China is a fragmented society. The fragmented society of China does not mean the society breaks into two parts; however, it may break into many parts in different dimensions and also within cities. Meanwhile, the society is still a dualistic society, as the most advanced area in China is connected to the market of developed countries and plays an important role in the world industrial market system. When the financial crisis happened, the economy in east coast areas of China was seriously damaged compared to the middle and western parts of China (Sun, 2004). According to Sun (2004), in the information society, joining the global market is a sword with two edges: on one side, it is a driving force from the outside which helps the country catch up to the pace of the whole world and also an important force to promote the whole society; the other edge is a force that may pull apart the society, since the force is not applied equally. As a result, the rich regions of China develop very fast, while the rural places struggle hard to catch up to the modernization process (Sun, 2004).

The opposite of the fragmented society is the plural society. Unlike the fragmented society, in the plural society the whole society is diverse and the social cohesion is well structured (Sun, 2004). Different groups are part of the society that contributes to the whole social system. The fragmented society may seem diverse if we just analyze it from the surface. The social disparity, the gap between the rich and the poor, different religions, values and cultures exist in both societies. However, the important difference between the two societies is that in the fragmented society, the social systems break into several isolated parts, and the subsystems also break into different parts, therefore the society is not a coherent whole system. In the fragmented society, the different areas and groups cannot integrate into the whole system; they live in the same country but in a different 'world'.

Compared to the first revolution, the industrial revolution accelerated technological innovation much faster and greatly improved the world (Toffler et al., 1981). However, how to understand modern China and which revolution it belongs to? As Toffler et al. (1981) said, the world has experienced three revolutions. Geographically, the east coast area such as Beijing, Shanghai, Jiangsu, Zhejiang, Wuhan, Shenzhen, etc. have witnessed new Science and Technology Parks springing up in the suburbs of these big cities. They produce goods and sell them to every corner of the world. The tons of goods that are made in China mostly originate from here. Currently, these regions are starting to focus on independent research and develop their own brands instead of just serving as Original Equipment Manufacturer (OEM) factories. These areas are the frontier areas of the booming economy and follow closely to the global economic trend.

To a certain extent, only the east coastal areas are participating in the third wave—a new revolution initiated by information technology and biotechnology. The rest of China remains backward in technique. In the middle and western parts of China, agriculture and heavy industry still contribute an important part of the whole revenue.

Compared to Western countries, the agricultural machinery in China is very backward, mainly manpower cultivation. The industries rely on labor resource instead of machines and the old, inefficient equipment is consuming considerable energy to maintain production. While the eastern area uses the computer to surf the Internet, the most laggard places probably still watch black and white television and have no concept of what the Internet is. As more and more youths study and work in cities, people who are left behind are mostly elderly, women and children. It is difficult for elderly and less educated people to learn technologies and the worst is the unavailability of physical access in rural areas. Therefore, in the fragmented society, the most advanced areas and laggard areas become disjointed regions of one country, breaking society into several parts.

In the developed countries, such as the United States, Germany and Canada, within different regions are different industrial levels. For instance, in the US, while Silicon Valley is designing the most advanced products, the peasants in the southern and middle area may still cultivate their plants. But the difference in the US and other developed countries is that the farmers may use technology and mechanized machinery to work, surf the Internet, check the market trends and use ICT to communicate with their customers just as the rich people do. The large-scale agricultural industrialization and the subsidies from the government guarantee that the purchasing capability of the farmers is similar to the other social classes. In Germany, the government uses taxes to balance the revenue gap among different strata of society so that even the lower classes can be integrated into the purchasing system. The peasants may work in the agricultural sector and can purchase goods as their urban counterparts can. However, in China, the new products or new digital devices are always purchased by members of the working class or the young generation. The farmer and middle-aged people stand in the marginal places of the information society.

Based on the above statement, this book assumes that contemporary China has entered into a quasi-information society instead of an information society. The eastern part, middle part and the western part of China represent the three revolutions respectively. The eastern part of China represents the third revolution, the middle of China witnesses the second revolution and the western part of China are still experiencing the first revolution, which leads to a transversally fragmented society. At the same time, ICT technology is developing very fast in China from the east part of China to the west part of China, even though the industrial level of east China and west China is different.

The other meaning of the fragmented society points to the different information the masses receive. As media becomes more widespread, the masses spend more time consuming it; as a result, people have less time to focus on the surrounding environment and less contact with the social system as a whole. In other words, they start to be less aware of what is happening in their society but more aware of what happened in another society. For instance, the individual ignores the events that happened in the local town, but knows many hot debates or striking news, which happened hundreds of miles away. They know every popular culture phenomenon and focus on Internet events, but know less about the neighborhoods where they live.

The attitude formation procedure of the audience largely depends on how they acquire information in their daily life. Many people heavily depend on the media for getting information, especially information about the economy and politics, and information relating to their living standard (Ball-Rokeach & DeFleur, 1976). As mentioned, the prior socialization decides which kinds of information the audiences choose to read. Here, it is necessary to address that the information tools they use also determine the information they receive. Such technical tools like television, computers and mobile phones are the main means to obtain information. Among them, different ones focus on different aspects of the society; for instance, the contents of computers are more diverse than the contents of television and the press. Therefore, people who tend to get information from the Internet have more chances to form diverse values (Gitlin, 1980; Herman & Chomsky, 2010).

For instance, the elderly people prefer to use television to get information and entertainment. The news they receive comes from China Central Television, which means the news content is more politically oriented. However, the young and middle-aged Chinese prefer to use mobile phones and computers to get information, which is more diverse so they can hear different opinions.

In the information society, the users possibly live in different societies if they use different devices, since the opinion of the society is largely influenced by the media.

The way people look at the world is affected by the pictures the media present them. The Internet breeds many hybrid languages, thousands of kinds of forums, and the users may use popular Internet jargon to identify themselves to show that they are caught up to the current trend or belong to specific groups. The diffusion of ICT shapes the society and generates different groups, which makes the society more fragmented. China represents interesting combinations of highly modernized regions, such as the eastern coast and big cities like Beijing, Shanghai, Tianjin, Chongqing; and non-modernized ones, for example, southwest China, especially those places inhabited by minorities.

## Conclusion

To better understand the background of modernization and the diffusion of ICT, I viewed China as a train. People inside the train can be categorized into two types, namely passengers who can use media and those who cannot. The aim of *hukou,* to restrict people from the rural areas from coming to work in the urban areas, is like the carriages of the train. However, urbanization and economic development open the doors between the carriages and mobilize the rural labor forces migrating to the urban areas.

By introducing the term fragmented society, I considered the eastern part of China as representing the third revolution; the middle, northern and western parts of China as witnessing the second revolution and still the first revolution, whereas the eastern and southern parts of China are more integrated with the information age. The wide gap between the rich and poor and the imbalanced development of different regions have shaped China into a fragmented society.

In the fragmented society, there are not only gaps among different regions, but also the ICT divides the masses into different groups, and users based on different media get different information. The movement of people who come from rural areas to work and study in the large cities updates the structure of China's labor force market. Passengers who use ICT are mostly influenced by the one-child policy, which means they are very special compared to the older generations.

The *hukou*, urbanization and the one-child policy are the most important variables influencing the fragmented society. By applying the train metaphor to China, it is easier to understand the background of the modernization process and the environment as ICT diffuses throughout China.

## Conclusion

The pervasive diffusion of ICT has dramatically changed China's economy, social system and the values of individuals as well as the relationship between government and the masses. However, the current social facts did not develop within a single day, rather step by step. Next, I will explain the development of media and analyze the mechanisms of the communication system behind the social reality. By knowing the development of media, it helps to understand the changes of media policies and how media adjust themselves to different social conditions.

# 2 The Eastern Media Philosophy

Free and responsible are two important factors when evaluating the performance of a country's media. Thus far, the censorship system in China is criticized by the West, and one of the criticisms is the media in China lacks freedom and liberty. However, the masses accept that the media in China lacks complete freedom, and it seems they don't have much need to fight for more liberty. To better understand China's media policy and the development of media, it is worthwhile to point out Eastern philosophy and how freedom and liberty were Western concepts brought to China by the Chinese intellectuals.

For most Asian countries that are influenced by Confucianism, it is important to protect a single company's interest, but it is more important to protect the whole interest of the country. The basic idea hides behind the typical Confucian's value that "survival for all" means not only survival for yourself but dedicating your life to the target of survival for the whole nation (Yin, 2003, 2008).

In 1956, three scholars who focused on communication study, namely Fred S. Siebert, Theodore Peterson and Wilbur Schramm, proposed four theories of the press. Afterwards, the four theories have become the main press models, which have dominated journalism education for decades. The four important theories presented logic and evolution behind the functioning of the whole world, especially the press in the Western world. According to the history and ideologies, they divided the development of the press into four stages and derived four theories, namely the authoritarian theory, the libertarian theory, the social responsibility theory and the Soviet Communist theory.

Since the advancement of the four theories, scholars have worked diligently in order to follow the development of the mass media. Many new theories arose in the late 20th century. Based on the four theories they proposed an elite power-group model, which characterized U.S. media as having a concentration of media outlets while being integrated with big business and government elites (power group). However, the fundamental problems with these media models are that the authors try to prescribe current systems rather than to describe social phenomena by using an empirical basis for inquiry (Ostini & Ostini, 2002). The development of media in China is influenced by Soviet Communist theory, and share similar characters of authoritarian theory.

According to the four theories of the press, during an authoritarian period groups are more important than individuals and the state can be regarded as a set and groups are elements. The individual as a member of society can only

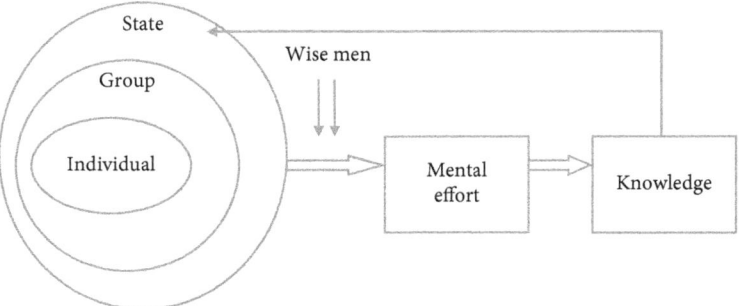

**Fig. 2-1:** Knowledge transfer in authoritarian theory adapted from the four theories of the press
Source: Author's compilation.

reach their goals within the boundaries of the state (Siebert & Schramm, 1956). The wise men have the capability to discover knowledge with their mental effort, which in turn will benefit the state and then the group and individuals. The wise men play important roles in the society (see Fig. 2-1). To the bottom, they inspire the masses, to the top, they give suggestions to the decision makers.

Liberty and freedom were regarded as a borrowed culture from the West, and still today these concepts are not fully adopted by China. The questions are why the media in China lack freedom and why the masses accept it? The *wise men* inspire the masses and give suggestions to the top, so, what were the initial values of the Chinese intellectuals?

## The Role of Intellectuals in Traditional Chinese Culture

### The Inital Values of the Chinese Intellectual

The concept of the intellectual is originally from Russia in the nineteenth century. The intellectuals, in general, have a strong critical spirit, especially with a moral critical sense, and are alienated from mainstream culture, holding their own independent beliefs (Xu, 2003, p. 3). The intellectual is also described as a group: "well-educated and critical of the traditional order; holds strong political ambition; and aspires to become a political leader or exert influence on the current policy system" (Xu, 2003, p. 2).

Yu Ying-shih claimed that intellectuals in Western culture are people who not only focus on their specific field but also care about issues of state, society and the whole world. Their consideration is based on beliefs, which encourage them to not only work for their personal interests, but instead, for the benefit of the whole world. Intellectuals have both intervention and detachment characteristics, which help them on the one hand interact with the outside, and on the other hand, keep their independent thinking, as I mentioned, that plays a role in looking out the window while other people are just sitting in the car. They live in the society but have to keep some distance from the real world, as they take a special role in the social division of labor to be the "social consciousness" of our society (Xu, 2003).

Taoism is a traditional philosophy that has had a profound influence on Chinese culture. Taoism believes in *yin-yang* (阴阳) and *wu wei* (无为), and encourages values such as naturalness, spontaneity, simplicity, and detachment from desires. Humans should live in harmony with the natural universe, follow its will, and not try to change or rebel against the will of nature; those are the essential values of *wu wei* and the basic doctrines of Taoism which can translate into "non-action". Those intellectuals influenced by Taoism appear detached, with less care about social issues and the future of the country. Taoism advocates spiritual freedom, the spirit is not restrained by the physical body, which is the supreme realm of Taoism; in this point of view, the basic idea is similar to the values of ancient Greek philosophies. Therefore, the Chinese intellectual prefers to maintain the social harmony, pursue freedom, and also be socially conscientious.

By the Spring and Autumn Periods,[12] *Shi* (士) had become a word which was only used by knowledgeable people. In this period, society had four hierarchy classes. The four classes were intellectuals (*shi* 士), peasants (*nong* 农), workers (*gong* 工) and merchants (*shang* 商). The *shi* class had priority over the other three classes as they were obligated to guide them and inspire the whole society. According to Confucius "*shi zhi yu dao* (士志于道)", "*shi buke bu hongyi, renzhong er daoyuan* (士不可以不弘毅，任重而道远)", pointing out that the intellectuals should focus on improving both their moral and ethical aspects, and at the same time, they should use their knowledge to guide politics and serve more people. The educated people should have a broad mind, cultivate a decent personality, watch their behavior and lead the masses. From this point of view, the role of the intellectual in traditional culture is similar to their role in Western culture, as *society's conscience*. There is a famous motto which says, "*xue'eryou ze shi* (学而优则仕)", meaning a good scholar can become an official. This sentence

---

12 It means *chun qiu shiqi* (春秋时期). It is a period in Chinese history from approximately 771 to 476 BC.

suggests that the traditional culture was aware of the close relationship between intellectuals and politics (Sun, 2012).

After having studied the intellectuals of China, Levenson (1964) held that there is no distance between Chinese intellectuals and the authorities, on the contrary, the Chinese intellectuals work as officers and provide service. Their role is to help the society think by using the identity of Confucianism. They never want to be ignored, neither in reality nor in the spiritual world (Levenson, 1964). This argument is proved by some articles and letters from intellectuals from 1880 to 1949. For example, when Gu Jiegang, an intellectual who chose academy as a career and life, investigated the middle part of China in 1931, he sadly expressed: "I wanted to focus and dedicate my life to knowledge, however, after I went to these places, what I have seen, and what I have experienced has led me to a road to save my country, help people and to do the things I can do. I should learn from Fan Zhongyan[13] who takes the destiny of his country as his own destiny (Ao, 2016)". Ding Wenjiang[14] said, "If we are not excellent intellectuals, who are? If we don't have the responsibility, who has? If we don't have the capability to change the situation, who does?" (Gao, 2000, p. 11). A similar voice was also expressed by Zhou Jianren, a famous writer who said, "Most Chinese people are illiterate, but we can understand characters so we have the opportunity and time to express our opinion. If we keep silent or talk little, the autocratic control will keep going, so we are the criminal in our soul, we will be accused if we do not represent the masses to speak" (Zhou, 1945, p. 3). "Fighting with the Japanese let me realize that my own life and the ordinary people's life have suffered day by day. I want to keep myself pure and only concentrate on knowledge, however, I cannot. For these two years, I abandoned books and now focus on political activities", said by Wen Yiduo (Sun, 2012, p. 31).

It can be seen with this introduction to the initial values of the Chinese intellectual, which are deeply influenced by collectivism and Confucianism, that they naturally care about the future of China. When Chinese intellectuals went to Western countries and tried to adopt new values, the individualism and new

---

13  Fan Zhongyan (范仲淹) (989-1052) was a prominent politician and literary figure in Song dynasty China. He was also a strategist and educator who executed many reforms and dedicated his life to the Song dynasty.
14  Ding Wenjiang (Chinese: 丁文江; March 20, 1887 – January 5, 1936), studied in Britain. After graduation, he returned to China and worked as a teacher. He was also a geologist, polymath, writer, politician, administrator, and social activist, most active after the establishment of the Republic of China in 1912.

ideas conflicted with their initial values which already existed inside, as a result, the initial values were reinforced instead of being changed.

## The Intellectual, Liberty and Freedom in China from 1880 to 1949

The spread of ICT and China's media policy have a close relation with liberty and freedom. To understand the development of these two terms, it is necessary to know what was the real thinking of the Chinese intellectual when they brought these values from the West; would they dedicate their knowledge to the country? The initial values they held were the traditional Chinese culture that was embedded in their self-awareness and composed of their self-experience. At the same time, some of the intellectuals had experiences of staying abroad which broadened their horizon as well as made them more patriotic compared to people who had never been abroad. They were always eager to dedicate their knowledge to change the destiny of China.

After the Opium War, the door to China was forced open. Many intellectuals who liked to "put their head out the train window" realized that the old empire of China was falling behind. Some of them studied abroad and adopted the values of Western culture and capitalism, and at the same time, foreigners' newspapers and broadcasts (most of them were run by missionaries) provided domestic intellectuals opportunity to get to know the world. The intellectuals regarded themselves as explorers who tried to find the future road of China. By enlightening the masses, the effort of reforms and revolutions finally succeeded in overthrowing the old social system.

Intellectuals who pursued freedom tended to adopt capitalism and Taoism. They obeyed destiny and held the value of "*wu wei* (无为)". The jobs of these intellectuals were similar, they were either teaching at schools or working in the press. As teachers, their main jobs were teaching, they buried themselves in knowledge year-by-year, cultivated students and collected news and sent it to the public. They hid in ivory towers, accompanied by books, what happened outside had nothing to do with them.

However, could they really remain calm when the country was in danger? The Qing dynasty chose to close the door to the outside and later on, the door to China was forced open instead of opening by itself. After the war between China and Japan, intellectuals could not hide in ivory towers as their consciences were questioned by their souls; they realized that they had a responsibility to inspire the masses to rise against the old Qing emperor and the enemies from Western countries, hence they used their pens as weapons to inspire the public.

Many articles or books have indicated the anxiety of the intellectuals. For instance, a famous writer Zhu Ziqing[15] said "Chinese Intellectuals hang between bureaucrats and the masses, at the top they cannot touch the sky, at the bottom they cannot touch the ground, which is a depressed living style, with many conflicts inside.[16] In the past, the intellectuals still could hide in the ivory tower, but now the so-called ivory tower is crashing gradually and changing into a crossroad. Therefore, intellectuals have to come outside and join the masses. People are tortured by the reality, however, if the reality stays like this, we will have to work hand in hand to break the reality[17]" (Sun, 2012, p. 14). Another famous author, Guo Moruo wrote that "It is useless to just stand on the crossroad forever; either stand on the right or stand on the left[18]" (Sun, 2012, p. 16). Hu Shih[19] said, "The first class talent should concentrate on the political field, the think tank should be people who work very efficiently, the first class talent should not choose the road of the industrial scientific field and leave mediocre people to control the country[20]" (Sun, 2012, p. 19).

The intellectuals were awoken by the complicated domestic situation that China found itself in due to its conflicts with enemy countries, therefore, they

---

15 He is a famous Chinese poet and essayist, November 22, 1898 – August 12, 1948. His earlier writing is more romantic. His later work is influenced by the war, becoming sad and radical. In 1937, the war between China and Japan started, when a student from Tsing Hua university asked his opinions, he said "国有难, 匹夫有责. 一个大时代就要来临, 文化人应该站出来" (Zhu, 1994). When translated into English it means, the country is in danger, intellectuals should stand in front and take action.

16 Originally in Chinese is: 中国知识阶级的文人吊在官僚和平民之间, 上不在天, 下不在田, 最是苦闷, 矛盾也最多." It means if the intellectuals in China lose both their relation to government and the masses then the role of intellectuals is useless.

17 Originally in Chinese is: 早些年他们还可以暂时躲在所谓象牙塔里. 到了现在这年头, 象牙塔已经变成了十字街头, 而且这塔已经开始 在拆卸了. 于是乎他们恐怕只有走出来, 走到人群里, 大家一同苦梅在 这活不下去的现状之中, 如果不满人意的现状老不改变, 大家恐怕忍不住要联合起来动手打破它的.."

18 Originally in Chinese is: "永远立在歧路口子上是没有用处的; 不是到左边来, 便是到右边去" (Guo, 1928).

19 Hu Shih, 胡适, who was born in Shanghai and studied at Cornell University in the United States. He was a Chinese philosopher, essayist and diplomat and is regarded as the most important person who contributed to Chinese liberalism. He played a key role in the May Fourth Movement and also, he was one of the leaders of China's New Culture Movement.

20 Originally in Chinese is: "应该有第一流人才集中的政治, 应该有效率最高的'智囊团'政治, 不应该让第一流的聪明才智都走到科学工业的路上去, 而剩下一批庸人去统治国" (Hu, 1998).

could not remain silent when faced with the situation that China might perish, albeit they tried to avoid getting involved in any political issues. Finally, they chose to participate in politics by establishing organizations, opening their own newspapers, calling for more people to join the fight, exploring new development routines for China and so on (Schwarcz, 1986).

Among these organizations which were established by intellectuals, the most successful was the Creation Society (*changzao she*创造社) set up in 1921. It played a crucial role in spreading new ideas and waking up the masses. The members of the Creation Society were mostly students who had studied in Japan, the representative members were Guo Moruo, Yu Dafu, Cheng Fangwu, etc. By establishing organizations and press, they spread new values to the masses and broadened their horizons. Since most of the members were students who studied in Japan, the difference in backgrounds distinguished them from the students who studied in American and European countries, since almost all of them chose the left wing side. On the one hand, they pursued freedom; on the other hand, they lacked an independent personality. In the May 30th Movement[21] most of them chose to become or support the left wing; the domestic trouble and foreign invasions finally pushed their attitudes from neutral to revolutionary. They chose to fight against the current government and the enemy countries.

Some intellectuals who adopted Western cultures, especially European and American cultures, tried to combine Western cultures with domestic reforms. Their roles were to be "good citizens" rather than "fighters". They were not against the current government but tried to convey their voice to it in a hope the government might adopt their values, which indicated they supported the legitimacy of the government, providing it would change its control strategy. They believed by using reason, law, and democracy they could reestablish the social system and actively participate in political issues and discussion.

After the May Thirtieth Movement (*wusan yundong*五卅运动), many innocent people died. The Chinese intellectuals criticized the masses that fought against the government as people without their own judgment who was only following the demonstrations. They regarded themselves as elite, by propounding concepts such as "planned politics", "new liberalism", "good government" and so on, they believed the role of the intellectual was to provide suggestions to

---

21 The May Thirtieth Moment, 五卅运动 in Chinese, was invoked by the killing of a child who worked in a Japanese company, which resulted in the labor anti-imperialist movement. A police officer in Shanghai opened fire and sparked national demonstrations and the death of many innocent people.

the government, and the smartest people should administrate and work for the country, not in science, and the government should have a mastermind alliance composed of the most intelligent people.

However, this duality made the intellectuals confused about their identity. Because they were not sure whether they, as intellectuals, should have an independent mind or belong to an elite class, which had to apply their wisdom and knowledge to coordinate conflict situations. As Hu Shih said, "Now I stand at a fork in a road, but late for three years; now I cannot restrain myself to not talk about politics, one of my feet stands on the East road, one of my feet stands on the West road, but my heart is still hesitating in the old road."[22] Ding Wenjiang said, "When I was young, I went to the most developed countries to study. I have been to Germany, Soviet Union……when it came time to leave, I asked myself, if I could choose, would I want to be a worker in the United Kingdom or the United States or an intellectual in the Soviet Union? I said without any doubt 'a geological technician in the Soviet Union'! If we have a Chinese style dictatorial government, we might have a chance to remain independent, otherwise either commit suicide or become obedient Japanese citizens."[23]

These words expressed that they intended to utilize freedom and democracy to inspire the government and save the people who were trapped in misery, but the result was they lost themselves as well. In the end, most of them worked for the government and served as the propagandist, their mild and hesitant attitudes guided them far away from their original beliefs.

The intellectuals were deeply influenced by traditional culture and believed the values of Confucius. Liang Shuming and Feng Youlan were part of the first generation of New Confucianists (*xin rujia*新儒家) and are good representatives of the movement (Chang, 1976). New Confucianism is an intellectual movement which is made up of intellectuals who insist on using values of Confucianism to solve social issues (D. A. Bell, 2010). They hold that the Western cultures and philosophies cannot be applied to the situation in China; only the wisdom of the traditional culture can help China find the right way. However, during this time,

---

22 The original text is: 我在这三岔路口, 也曾迟回了三年; 我现在忍着心肠来谈政治, 一只脚已踏出了东街, 一只脚还踏在西街, 我的头还回望着那原来的老路上" (Hu, 2013).

23 Originally text is: "我少年曾在民治政治最发达的国家读过书的. 一年以前 我到德意志苏俄参观过. 我离开苏俄的时候, 在火车里我曾问我自己: 假如我能够自由选择, 我还是要作英美的工人, 或是苏俄的知识阶级?我毫不迟疑地答道,'苏俄的地质技师'! 在今日的中国新式的独裁如果能够发生, 或许我们还可以保存我们的独立. 要不然只好自杀和作日本帝国的顺民了" (Ding, 1935).

the traditional (Ao, 2016) culture had become an obstacle to social-economic development and the social order was too stable and compressed to allow space for new ideas and values. During this period, China had suffered both inside despotic control and outside invasion by foreigners, as a result, religions such as Buddhism had become prevalent as they could provide spiritual comfort when real life had become torturous. The intellectuals focused on Buddhism as well, combining Buddhism, Taoism, Christianity (some intellectuals studied in Western countries and became Christian) and other religions and culture into their values, as a result, New Confucianism is characterized by a mix of cultures such as Confucianism, Taoism, and Buddhism (D. A. Bell, 2010).

In order to inspire the masses and fight for the future of China, the intellectuals served as special writers in the government to help write news and reports. However, some intellectuals chose the opposite way—they hid in the towers of ivory to remain isolated from the outside. They believed in *wu wei* and bore the torture of reality, they focused on their own heart and reflected on the nature of Chinese people and Chinese culture. (Alitto, 1986; Cheng & Bunnin, 2008; Tu, 2001). Liang Shuming held that Chinese do not understand the real meaning of the concept 'country', they only know peace. If the government didn't intervene in their peaceful life, they wouldn't care about the future of the country. He said he agreed and also loved a peaceful life ((Liang, 2011).

The values of the intellectuals during this time were impacted by both Western culture and traditional culture (Tan, 2008). The triumph of Western culture is established on the high living standard of the Western countries, which gives those living in the third world an illusion that Western culture has superior values, that only Western culture can lead to the right way and help China find the road to prosperity. This illusion also made the intellectuals who believed that only Chinese culture could save China start to doubt their beliefs; they questioned the traditional Chinese culture, and questioned themselves as adopters of the culture. This ambiguous feeling is familiar, as previously described concerning the people of the indigenous villages faced with the tourism economy and modernization, they also questioned their local culture. In this situation, will Western values be able to lead China to a bright future? How were Western values brought to China? Next, I will introduce the history of how liberty and freedom were brought to China.

## The Introduction of Liberty and Freedom to China

Taoism has played an important role in shaping traditional Chinese cultures. *Wu wei* (无为) and *yin-yang* (阴阳) are the core values of Taoism. It encourages inhabitants not to rebel against the will of nature; they should be obedient and trust in their destiny. Confucianism promotes order, which can help the society avoid a chaotic state. Freedom and liberty have existed in China for a long time, but did not develop into mature values; respecting the order and social scale are still dominate values which highlight the role of Confucianism (Kohn, 2001; Maspero, 1981).

Confucius, Mencius and Zhuang Zhou believed people have destinies. Individuals' hearts can have freedom, provided they accept their destinies. Zhuang Zhou stressed that the individual's will and mind should be free from the material world. Confucius and Mencius held that the nature of human beings is good and they can achieve freedom in spirituality. Xunzi developed this value and combined it with nature, heart and freedom. He believed human beings are essentially free because of their heart, with which they can decide what kind of orders they accept. The outside forces can keep them silent, but cannot change their will. The reason that the heart feels bad is that it cannot calm down; only if the heart is quiet, can it have freedom. The freedom heart is based on reason and rationality. Afterwards, two important people, namely, Wang Shouren during the Song dynasty and Li Zhi during the Ming dynasty developed the idea of freedom in a different way. They encouraged cultivation of an individual's own personality, and held that there is no boundary between truth and falseness. Afterward, in the Qing dynasty, Wang Fuzhi, Gu Yanwu, Huang Zongxi, etc. wrote from the perspective of an individual's right to rethink freedom.

However, for a long time, freedom and liberty existed in the traditional literature as a concept that is in opposition to the concept of order, which is an important value in Confucianism. It became a real value when intellectuals such as Yan Fu and Liang Qichao studied in Western countries and introduced these concepts to China. After these new concepts were introduced to China, the values served as weapons to fight against the feudal emperor and autocratic government, and the Chinese intellectuals tried to utilize these concepts and applied them to China in the early 20th century (Liu & Chen, 1988b; Mote & Rogers, 1988; Munro, 1969).

When liberty was introduced to China, it was shaped by the values of translators. One of the important translators was Yan Fu. He was not only familiar with traditional Chinese culture but also knew Western culture very well. John Stuart Mill's *On Liberty* was translated by Yan Fu in 1903, and he imported the concept of liberty systematically to China. However, his book *On Liberty* was not precisely translated according to the original version; it was the translator's ideas combined with Mill's ideas (Huang, 2003; Huang, 2008).

Yan Fu's values were changing during his life, but there are three main values we can observe from his works and these values are characterized by three different periods. From 1895 to 1900, he started to form his own ideas on liberty, at the same time, he advocated liberty and freedom by translating books and publishing articles in newspapers to spread liberty. From 1901 to 1910, he reflected on himself, and the books he translated are mixed with his own ideas, which differ from the original writer. From 1911 to 1921, he tried to find the value of freedom from traditional Chinese culture and combine it with the liberty concept from Western books, in order to create a new concept that mixed traditional Chinese culture within a liberalism context. Therefore, when liberty was first introduced to China, it was conservative *and* radical (Huang, 2003; Huang, 2008; Lackner, Amelung, & Kurtz, 2001).

According to Yan Fu, the concept of freedom and liberty do not exist in the traditional values. The alternation of new and old dynasties and autocratic control resulted in the Chinese elites lacking the courage to develop the value of freedom. The inhabitants got used to bearing the miserable life. For a long time, the value of "the gun always kills the first bird (*qiang da chutou niao*抢打出头鸟)" has been deeply embedded into the Chinese mind—they do not want to be special or show their talents. They prefer to be ordinary people; no one wants to be the first bird who tries to escape the cage or brush that might be killed.

In terms of democracy, the early intellectuals who brought Western culture to China believed that democracy is the tool of freedom (Chao & Myers, 1998). They thought democracy is generated by freedom. Without freedom or liberty, democracy does not exist. Yan Fu believed liberty is the essential idea, and democracy is a tool that reflects liberty. However, Yan Fu was also a follower of Darwinism. He believed Western people were physically better than Chinese, and the colonial Chinese society intensified this belief, which he believed could only be overcome by solidarity, collectivism and the unity of the majority— together they could help China become a rich country and eliminate the dangerous situation. In the later works of Yan Fu, he held that next to freedom and liberty the most important tasks for China were to save the country and help China become a prosperous country (Huang, 2003).

Liang Qichao is another intellectual who had an important role in bringing Western values to China. When discussing the destiny of China, Liang Qichao claimed that the traditional Chinese culture had no definition of freedom, however, if China could not understand the value that civilians were not just people, but people with freedom, the inhabits in China were just slaves and not an independent nation. If the Chinese could not change this value, no matter what they did, e.g., introduce new technology, change the political structure and so on, all the efforts were just in vain (Wu, 2006). Spencer's social organism concept, that the education level of one country decides its industrial level, had inspired Liang Qichao. He realized that compared to the Western countries, lack of freedom resulted in China being in a dependent situation. For the sake of the future of China, it was necessary to re-educate the citizens and spread the value of freedom. Based on the values of Liang Qichao, the spread of freedom, democracy and literature were necessary when considering the future of China (Wong, 1992).

However, during that time, critics argued that the concepts of freedom and liberty do not exist as values, but as tools to change the people's mind in order to realize the political ambitions of the intellectuals (Ames & Hall, 2015; Held, 1995). The inherent shortage of freedom and liberty was like an incomplete, weak seed that could not grow strong in the earth of China. By the time the Democratic League was dismissed in 1947, almost forty years of effort had not changed the status of freedom and liberty, but the traditional values had successfully integrated these Western values with new meaning (Ames & Hall, 2015). It seems the intervention of modernization could not make traditional culture fade away, on the contrary, traditional culture made a strong impact on modern culture.

By introducing the early stages of freedom and libertarian thought in China, I have shown how the Chinese intellectuals stood at the crossroads, finally choosing to participate in politics and combine the values of freedom and liberty to their political opinions. They believed that to be a politician or stand by the side of a politician was the best way to have their values adopted by the public and consequently influence the whole society. The intellectuals who tried to introduce more modern but uniquely Chinese values and philosophy to China seemed to have gotten lost and hesitated between the ideal and reality. And still, the topic of freedom and liberty is serious enough that magazines, newspapers, books, broadcasts, ICT and political systems have to carefully deal with the boundary between the masses needs and the government's wants.

## New Concepts without Real Meaning

After the new concepts of liberty, freedom and democracy were brought to China, many intellectuals, after comparing traditional and Western culture, found that in the Chinese dictionary there were no words that had meanings similar to freedom and liberty. In Chinese, freedom means *ziyou* (自由), *zi* means oneself, *you* means reason. The meaning of *ziyou* is not the same as freedom, which was brought from the West. Therefore, the intellectuals tried to build new meanings for the introduced concepts to make them fit in the context of China. From one side, they reflected about the traditional values; from the other side, they enriched the traditional values by selecting the elements of the Western values that could fit into the traditional culture, for example, Sun Yat-sen.

Sun Yat-sen is one of the greatest leaders in the history of modern China. His political philosophy about nationalism, democracy and people's livelihood (Three Principles of the People) has an important influence on modern China. Regarding freedom, he had spectacular views as he claimed that in the past, China had too much freedom. As in Western countries, the degree of freedom of the individual was decided by religions and states. However, in China, if individuals paid a tax to the emperor, then they finished their responsibilities. China has no religions. Whether Confucianism, Taoism or Buddhism, these are just value systems, not real religions. Without the restriction of religions, the Chinese lack a sense of nation. The collectivism might exist at the level of family, but at the level of the state, the majority of people do not really care (Cui, 2009).[24] When China was invaded by Western countries, the difficult problem was, as Sun Yat-sen argued, how to unite the loose sand together to fight against the enemy and the old social system (Schiffrin, 1968). Sun Yat-sen believed the Chinese people had too much freedom in the past and that laws or rules were hard to apply in China. For a country that is without law, rules, or religions, if freedom and liberty are introduced, the values would lose their original meaning (Hsu & Sun, 1933; Sun. Y, 2012).

Reality pushed the intellectuals to combine Western values and the destiny of China together. Hence, the intellectuals created a new concept called "National Freedom". It meant China should fight for becoming an independent state, should win the respect of the world, and should be a strong country able to decide its own destiny (Ip, 1994; Mao, 1945; Schwarcz, 1986). The Chinese intellectual found that in traditional Chinese culture, the social stratification was very strict, and as a result, ordinary people did not have the opportunity to protect their own rights, consequently, the sense of responsibility and obligation were also weak. The National Freedom concept endowed the new meaning of freedom and liberty in the context of traditional culture, which meant the individual had freedom, provided the nation was free.

The ancient Chinese empire believed it was the center of the world, which in Chinese is called *zhongguo*, *zhong* means middle, guo means country. When foreigners forced China to open the door, the conflict between traditional culture

---

24  More information can be found in http://blog.boxun.com/hero/sarmin/6_6.shtml. Accessed July 26, 2015.

and Western culture pushed many people to reflect on China and the traditional culture. Intellectuals, as the group that stood in front, were deeply influenced by traditional culture and the reality pushed them to learn new values from developed countries and utilize these values to free the country.

The intellectuals imported Western values to China between the 1920s and 1930s; during that time, China had experienced domestic war and economic crisis (Chao & Myers, 1998; Huang, 2008). The intellectuals found it was difficult to transplant new values to China as the environment did not fit. However, the prosperity of the Western countries gave an illusion to the intellectuals that freedom and liberty could save China. As a result, the new values were endowed with a destiny to save China, which means after the concepts of freedom and democracy were transplanted to China, they had already lost their original meaning.

In 1937, Mao Zedong wrote an article called *Combat Liberalism*. In his article, he argued that the conflicts of different values and peace without rules will jeopardize the solidarity of China, and liberalism will result as the consequence (Mao, 1945; Mao, 1983). Mao's understanding of freedom and liberty has a direct influence on the ideology of the Communist party. After analyzing the original meaning of liberty and the relationship between liberalism and Marxism, Mao concluded that the concept of liberty reflected the selfishness of capitalists and regarded it as opportunism, which he believed, conflicts with Marxism (Benton, 1984; Ip, 1994). As a result, when the communist became the ruling party, freedom and liberty became ambiguous concepts. Consequently, during Mao's period, both the media and the central government tried to avoid the topic of freedom and liberty (Schwarcz, 1986).

In the 1950s, the debate between New Confucianism and Liberalism set off a wave of reflection on traditional culture and new Western values. The New Confucianism was represented by Yin Haiguang and Mou Zongsan, the arguments between them existed for many years, and climaxed with mutual hatred and the refusal to speak to each other until the end of their lives (Wang, 2009; Wang, 2009). The core point they argued about was whether the traditional culture in contemporary China is useless or not, whether the traditional culture can support the values of individuals when the situation is changed. They spread their ideas via the media, the press was their battle place and pens were their weapons. This debate ended in a compromise where the liberalists admitted the value of traditional culture and that Western culture could survive in China only in combination with it (Yung, 2015).

When explaining the reasons why Chinese people who are influenced by Confucianism tend to refuse liberty, Yin Huaiguang gave a special answer. He held that few Eastern people have experienced freedom (Sun, 2012). The freedom that is like a heap of loose sand is not real freedom. Liberty and freedom became ambiguous concepts for the masses as well as for the intellectuals. The masses did not know what real freedom is even though they yearned for freedom. Some people who were afraid of freedom explained freedom as an evil word and thought it equal to indiscipline, not following rules and so on (Wu, 2006). For non-democratic countries or those with less freedom, revolutionists always use freedom against the government. However, when the government is at risk or faced with break up, absent of any form of political authority, anarchy will follow. As a result, tyranny will dominate the situation, then the vicious circle will start again and again, which implies that freedom and liberty are dead (Ames & Hall, 2015).

Freedom and democracy not only lost their original meaning from the beginning, but also could not develop. Why couldn't the freedom revolution fully develop in China? In Western culture, the idea of freedom had developed over a long time and each step of the experience could be observed. But in China, foreign military invasions and domestic war destroyed the environment that might generate freedom and democracy. When intellectuals brought ideas of freedom and democracy to China, these two concepts were mixed, and lost their original meaning since they were selected from different types of liberalism and democracies, e.g., some intellectuals accepted America's idea of liberalism whereas others adopted European liberalism, which has many branches of its own. As a result, when freedom was brought to China, its meanings were only partly adopted.

Modern Chinese history is a violent and revolutionary history. This difficult situation pushed intellectuals to join the political parade (Grieder, 1970). After the fight with the Japanese, the domestic war began; some of the intellectuals chose to support the communists, and some of them chose to stay with the nationalists. At this crossroad, intellectuals abandoned their freedom briefly and stood in the political arena. In 1941, the China Democratic League was established. It only existed for six years and broke up in 1947. However, the disintegration of the Democratic League indicated that the bubble of intellectuals who wanted to use democracy and freedom to change China was broken. Grieder (1970) attributed the failure of liberalism to the fact that life in China was molded by violence, which opposes the nature of freedom, as liberalism is based on reason and freedom (Grieder, 1970).

Culture is like plants and animals which can survive very well in the place where they adapt to and the season in which they fit; values and beliefs are difficult to extend beyond their time and can only be fully understood in the context of the time and space in which they live. In biology, each species experiences different evolution. If a plant species is transplanted to another place, it needs self-adjustment in order to adapt to the new environment. It may develop new functions during the self-adjustment procedure, and such adjustment needs to be observed continuously.

However, plants and animals can be moved to new environments as whole units, culture cannot transplant as a whole unit. In the 17th century, the potato was imported to China and became an important crop that saved many lives at that time. When the potato was imported to China, it was brought as a whole unit – potato, not just seeds. Unlike the potato, cultures or values cannot be imported as a complete unit since individuals are carriers for the culture that each individual adopted based on their personal experiences. In this case, when culture or values are transferred from one place to another place, they are usually not transferred as a whole unit. During the spreading process, the individuals' values are the initial values, which means they selectively choose parts of the new values that fit their initial values.

## Conclusion

The intellectuals in China are influenced by Confucianism and traditional culture. Using knowledge to dedicate oneself to the interest of the country is an idea embedded in the values of Chinese intellectuals. Government plays a paternalistic role in the Confucian hierarchical system, which means the government has an obligation to make decisions and individuals have the duty to follow these rules and decisions. Thus intellectuals, as the brain of society, in order to influence it and spread their knowledge and values to the masses, should first influence the government via the media. Since the media serves the state, therefore, the knowledge that intellectuals want to spread via media shall not be against the government.

The initial value decides what kind of value they adopt and what kind of value they abandon. When the Chinese intellectuals went abroad to study, first they experienced culture shock. The Eastern culture they grew up with was embedded in their values and conflicted with the new Western values they encountered, which is one of the important reasons attributed to their difficulty in adopting the whole concept. Besides, transferring the culture is different than transferring a plant, which can be transplanted as a whole unit. This caused them to reflect and attempt to incorporate the new Western values without internal conflict. Thus, the concepts of freedom and democracy were adopted in a way that was cohesive with their Eastern philosophy.

# 3 The Two Stages of Development of the Media in China after 1949

The importance of knowledge as a basis for social power has been noted by a number of scholars, but less focus on the fact that control of knowledge is a central task for maintenance of power. The knowledge industry accounts for more than a fourth of the gross national product and attests to the social importance of the demand for knowledge production and distribution (Kerr, 2001).

Mass media is receiving increased attention in strategic management because it might be an intangible resource leading to the sustained competitive advantage. Politicians like to use mass media to overcome a hostile environment by propagating value to win societal stability. In order to keep social stability, people who control media will select the best media to fit the social condition; hence, individuals who mostly get information from mass media are affected as well.

The social condition can be distinguished into two states: social stability and social disorder. A stable society is at the point where the subsystems are in a balanced state, the different subsystems are cooperating and a social equilibrium has been reached. Social stability is regarded as the foundation of a harmonious society. The opposite of social stability is social disorder, which means the balance is broken up and society is in chaos (Birner & Ege, 1999; Gilboa & Matsui, 1991).

The stability of society, which is the root of political economy and sociology, inspired Adam Smith to develop the idea of the invisible hand. If society is regarded as a whole system, then the social system is viewed as a series of interrelated subsystems. The main functions of these subsystems are to generalize, disseminate and assimilate information to affect further control as a means to an end or as an end in itself (DeFleur, 1966). ICT works as a subsystem that is responsible to transfer and control information, the most important capability is it can communicate with other subsystems, acting as an information hub.

The media can be differentiated into four types: listening mass media, such as radio; reading media, for instance, newspaper and magazine; watching media, like television; and new media (mainly characterized by ICT); for example, computers and smartphones. The old media is comprised of newspapers, magazines, fixed-phones, televisions and so on. ICT includes fixed-phones, cell phones, television sets and the Internet (with personal computers), etc. The ICTs focus is digital computing and telecommunications technologies (Davison, 2016; Haddon, 2004; Heeks, 2002; Sahay, 2016; Silverstone & Haddon, 1996).

In the course of the spread of information and communication technologies such as smartphones and different Internet surfing devices, which are all endowed with capabilities of connecting to the Internet, the term ICT becomes a popular concept which points to technologies capable of both processing and communicating information. As ICT diffuses, online news replaces newspapers, while e-books and e-magazines push traditional bookstores and magazine agencies to modify their strategies to adapt to the information age.

With new media, the audience is not just an information receiver but can interact via these media to become an information publisher. As I have introduced the development of radio as well as how the government gave it preference, it will be shown that the development of ICT relates to politics, and the body politic will have to adjust their propaganda strategy according to the function of the media.

As O'Shaughnessy and Stadler said, *"Traditionally, two models have been suggested as ways of understanding the relationship between media and society. The first suggests that the media reflect the realities, values, and norms of a society. Thus if we want to study a society we could turn to its media—its films, novels, television series, and popular stories. They will reflect to us what people feel and think how they behave, and so on. The media act as a mirror of society, or a 'window on the world,' which can be used as a resource to understand the society. The second model suggests that the media affect how people think, believe, and behave. The media construct our values for us and have a direct effect on our actions."* (O'Shaughnessy & Stadler, 2012, p. 24) Based on the relationship between ICT and social conditions, I will design a model to explain the mechanism of ICT's interaction with society.

The model (Fig. 3-1) contains an outside and an inside circle. The outside circle explains the relationship between media, individual, social condition and government. The inside circle illustrates that during the different social condition, namely social stability and social disorder period, the media (we media, television, radio, fixed-phone and the press) have different roles in the communication system.

## The Outside Circle: Interaction of the Media and Society

The outside circle is like a societal adjustment system. In the information society, the relation between government and individual is not a linear relationship, however, it is more like an indirect relationship. Via the media, the government and individuals communicate with each other. The society is made up of each individual, which means the state of individuals decides the condition of the society. However, the values of individuals are largely shaped by the media, and

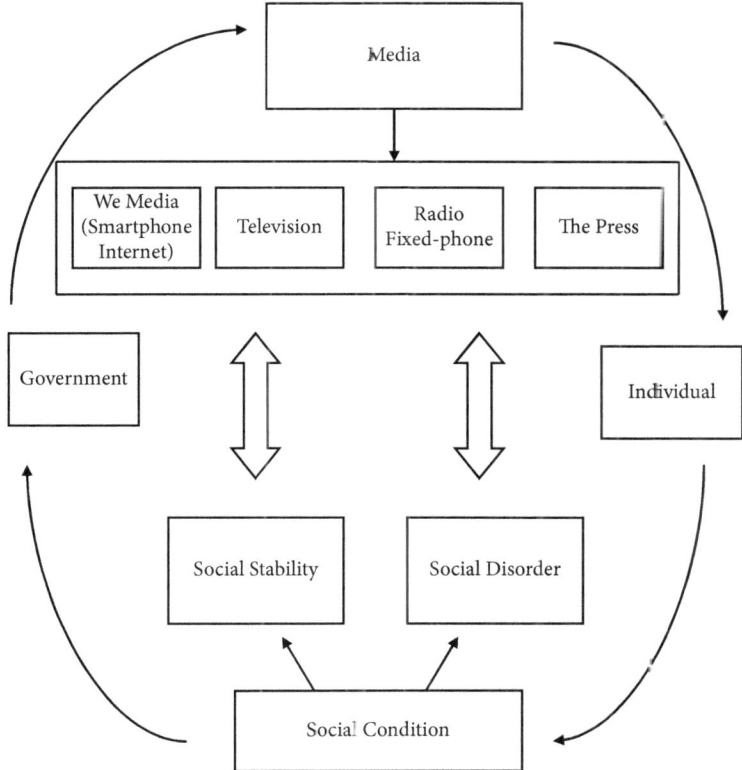

**Fig. 3-1:** Media and social condition model
Source: Author's compilation.

media reflect the social condition (see Fig. 3-2). Consequently, the social condition influences the government's policy and strategy for administration. This is the outside circle.

Society itself is composed of a series of social experiments. A new policy or reform is successful or fails depending on the coordination of different subsystems. Before a new policy or reform is implemented, the government likes to use the mass media to make tests. In addition, people or a group who control the mass media can observe the social trend to estimate the future. In terms of China, the media are controlled by the government, by collecting topics or trends that are interesting to the public; the government can guide directions of popular topics and make decisions.

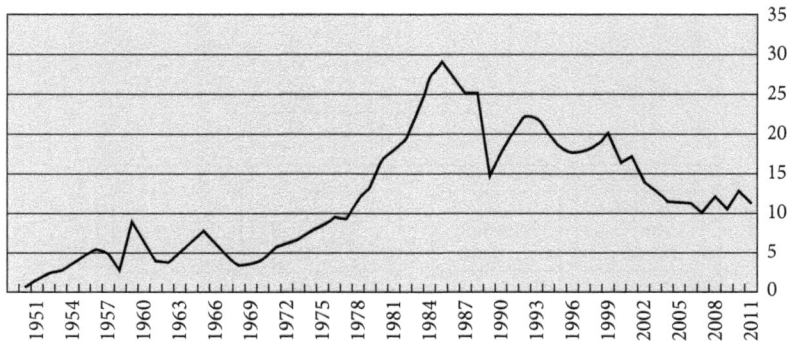

Fig. 3-2: Annual average number of newspaper and magazine subscribers per 100 persons

Source: National Bureau of Statistics of China (NBSC, 2016).

Note: Figure 3-2 shows the annual average number of newspaper and magazine subscribers per 100 persons. It suggests that the peak and bottom points of annual issues of newspapers and magazines are closely related to the social condition. The lowest point happened in 1989, which connects to the Tiananmen Square protests of 1989. The publishing systems are mainly controlled by the government and less affected by the needs of the market. From the whole trend, it also suggests that the annual issues of newspapers decreased.

## The Inside Circle: Social Condition and the Media

The inside circle is more complicated. From the media and social condition model (Fig. 3-1), four types of media can be distinguished according to their main functions. The social condition that is mainly composed of social stability and social disorder both influence the government, afterwards it chooses what kind of media to spread their propaganda (Fig. 3-3). Individuals in the society determine the social condition but are also affected by the media, as the media are the information-gathering system as well as an important part of the public sphere, allowing social members to contact each other and organize activities.

During the past few decades, China has experienced both economic and societal change, transforming from a centrally planned economic system to a market-based capitalist economic system. Similar to the economic change, the media adjusts themselves to the speed of economic system change. In 1978, Deng Xiaoping started the economic reform process, setting off a chain reaction in the economy and social structure, as well as in the communication system. When analyzing the mass media in China, I will use Mao's period and the post-Mao period as two dimensions to stress the two different periods in China.

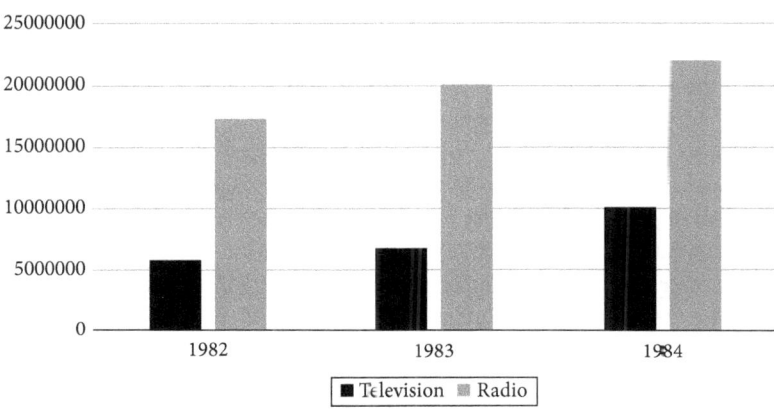

**Fig. 3-3:** Adoption number of television and radio from 1982 to 1984
Source: MIIT (2016).
Note: From 1982 to 1984, the adoption rate of radio is much more than television. Compared to the television, the radio is cheaper. The function of radio is favored by the government and fits the social condition.

## Mao's Period—Slogans and Social Mobilization

The media in Mao's period are closely combined with social mobilization. When Mao became the leader in China, he deeply believed the success of communism is established on the mobilization of the masses. Either during the second Sino-Japanese War or a domestic war, he believed only the masses can decide whether a war will be won or lost. Mao held that the revolution belongs to the masses; only by mobilizing them can the communists stay aligned with the masses and defeat the enemy.[25] As Mao had stressed that the press was the important tool to spread and encourage the masses and that the communists should utilize the mass media to serve for them, as a result, when new China was established the communists adopted Mao's opinion that every movement should depend on the masses.

---

25  Originally text is "依靠民众则一切困难能够克服，任何强敌能够战胜，离开民众则将一事无成" (Mao, 1938). It means depend on the masses and we can overcome any difficulty, and can defeat strong enemies. Without the masses, we will lose everything. "只有坚决地广泛地发动全体的民众，方能在战争的一切需要上给以无穷无尽的供给" (Mao, 1934). It means only by widely mobilizing the masses, can we furnish the needs of the war.

The central government believed slogans were important to mobilize the masses to adopt the ideology. They created many slogans for the mass media and printed them on the walls of houses to make sure they were seen by everyone. Slogans can be divided into specific slogans and vague slogans, for instance, One-child policy is a specific slogan, under this big slogan there are many small specialized slogans, such as *"Fewer children, better life"*. Under the vague slogan, there are also many small slogans, like *"Transfer positive energy"*. The crucial value of the slogan in the press can trace back to Mao's period. When Mao investigated the movement of farmers in Hunan Province, he said "A simple slogan and cartoon pictures can educate the masses with good efficiency. With different slogans, the masses are educated in the political school and only positive slogans together with positive attitudes can mobilize the masses" (Mao, 1991, p. 102). Indeed, the brief and simple slogan had immense power to mobilize the masses. During the Great Leap Forward and the Cultural Revolution, the masses were filled with slogans. Still, the media in China like to use slogans to spread policy and propagate the ideology.

In Mao's period, the central government made a decision, and then the decision was transferred by meetings and documents to the sub-governments. Afterwards, sub-governments communicated via media the resolution to the masses. For example, in the Great Leap Forward the government preferred to combine individuals into different groups, namely production teams (*shengchan dui*生产队) in agriculture and work units in the industry. The information transmission is like the graphic in fig. 3-4.

Rogers (2010) made the point that government influences the innovation and diffusion process through regulation, such as policies that can have substantial effects. In Mao's time, both traditional and new media are diffused under the frame of government strategy and policy. The government attached importance to the mobilization of the masses while the development of the media was influenced by the circumstances of the society.

## Press

The newspaper system was influenced by the Soviet Union when China was established. It was a specialized system, with different groups getting a different newspaper (business newspaper, student newspaper, peasant newspaper, military newspaper, financial newspaper, etc.). The press in China is a complicated and hierarchical system which is based on the role of media as a servant for the ruling party.

Under the hierarchy system, the press and the media are endowed with authority to supervise local governments and organs that are at the same level

**Fig. 3-4:** News transmission in Mao's period
Source: Author's compilation

and below the authority level. From central government, to provinces, cities, and towns, every administrative unit has its own newspaper agencies and they have administrative authority to supervise the secondary governments and are responsible for propaganda and news released by the local government. Therefore, to some extent, the newspaper agencies are set up by organs and local governments in order to supervise each other. More precisely, in the hierarchy press system, the newspaper agencies of provinces can censor the press of province, towns and cities; likewise, the press of a town can censor towns and villages below the authority level. The hierarchy press system is composed of several thousands of newspapers. Almost every local government, except county and village, has party organs responsible for publishing news and conveying information from the central and local governments to the masses.

The Secretary of the Party in charge of the press and from top-down, every administration had press agencies, and every day the communists worked together with the masses to discuss the newspaper, including headlines, slogans, role models, etc. and fed the comments back to the top. As a result, the task of newspaper discussion became so heavy that other important work was delayed.[26] During this social disorder period, the content in the press was all positive news, the media tried to describe an improving, pure and dynamic society to make the masses believe the government chose the right road for them. For a long time, it is a tradition that the media in China is forbidden to report the negative news, but instead should mold a fantasy picture of rosy scenery for the

---

26  The statement is according to the investigation from Anhui province in March 2015.

masses to convince them that they live in a world like the mass media describes (Swanson, 1996).

Before 1978, the newspapers in China mainly focused on political news. From the central government to local governments, all different administrations had their organ newspaper. Among these organ newspapers, the People's Daily (*Renmin ribao*人民日报) was one of the largest newspaper agencies, which represented the opinions of the central government (Yao, 2001).

The press played a very important role in the Great Leap Forward, since the People's Daily was nearly a half-leader during that time (Swanson, 1996). The slogan of the Great Leap Forward was first introduced by the People's Daily. When Mao Zedong wrote a letter to Liu Jianxun and Wei Guoqing, he said "the newspaper for the province is crucial, it leads to the work of the province, it encourages the local people, it inspires the masses, it mobilizes the masses and so on (Mao, 1983, p. 202)."

The Great Leap Forward and the Cultural Revolution are examples of government mobilization of the masses with the press. In order to better serve for the Great Leap Forward, the circulation of the press was increased and the content of the press was very homogenous. During Mao's period, by using the force of the masses and the press, the government achieved economic improvement and fought in a hostile environment.

**Radio**

Since the Central Broadcast Station was established in Nanjing on August 1st in 1928, it was controlled by the Nationalist Government. The programs contained news and speeches, as well as international news. In order to censor local broadcast stations, the Nationalist Government established a committee to supervise newspapers, television and radio. In 1940, the Communist Party established Xinhua Radio Broadcast Station in Yan'an called XNCR, which belonged to the Xinhua Agency. The broadcast used three languages: Chinese, English and Japanese, to transmit messages to domestic audiences and the outside world. After the People's Republic of China was established the XNCR changed its name to China National Radio (Zhao, 2004). From the 1950s to the 1990s, almost every village had its own radio station, and listening to the radio was the most common way for Chinese to receive information.

Similar to the press hierarchy system, the early broadcast system consisted of three levels—central government, provinces and city. Each hierarchy had a broadcast station. The contents of the broadcast can be categorized into several types, namely Korean War, land reform, minority integration, economics, politics, etc. Moreover, radio was used to communicate between the communist and nationalist parties. The communist government produced radio programs focused on Taiwan, as Taiwan was controlled by the nationalist party, and likewise, the Taiwan government broadcast programs intended for China as the two parties communicated with each other and propagated their ideologies to audiences (Ai, 2002).

During the Great Leap Forward and the Cultural Revolution, the radio developed very fast. The Great Leap Forward started in 1958, afterward, the diffusion of radio in the whole country lapsed into an abnormal routine. According to the Ministry of Industry and Information Technology of the People's Republic of China,[27] between 1958 and 1960, transmitter-receivers increased from 91 to 137 and then dropped to 89 in 1963. Social disorder stimulates the diffusion of media. From 1957 to 1960, the number of broadcast stations increased from 1698 to 2404 and loudspeakers grew from 0.94 million to 6.04 million. The number of radios grew to 1.58 million in 1960, which is 3.5 times more than in 1957. Television stations increased from 61 in 1957 to 137 in 1960. A case to illustrate the quick diffusion of media in the Great Leap Forward period is in Fujian province, where in 1958 the diffusion of broadcast stations and loudspeakers were 60 % and 80 % respectively, however, two years later, this number rocketed to 80 % and 100 % respectively. Anywhere with electricity would have a loudspeaker (Xu, 2004).

The radio and television programs spent a lot of effort on forged heroes (*shudianxing*树典型) and the famous such as "Lei Feng (雷锋)", "Wang Jinxi (王进喜)", "Jiao Yulu (焦裕禄)", "Deng Jiaxian (邓稼先)", etc.—these heroes' stories helped the central government spread their political values, and their stories were easily adopted by the masses (Zhao, 2004). Presently, with the widespread of ICT, the technological obstacles make the old generations gradually lose their voice in the new media and the new generations start to question whether these heroes existed or not. Regarding these issues, this book will return to them in the section on social memory.

When the People's Republic of China was established, the legitimacy of the communist party as a ruling party was not recognized by many countries. Moreover, Western countries committed an economic and military blockade to China. Mao Zedong realized it was necessary to implement military control of the media. The central government categorized the radio programs from Taiwan and

---

27 See http://www.miit.gov.cn/n1146312/n1146904/n1648372/index.html. Accessed June 25, 2015.

Western countries as enemy programs and tried to use broadcast waves to disrupt or block their programs. During the Cultural Revolution (1966–1976), the media was controlled by the military, invoking chaos in the broadcast industry. As the rebel group occupied the central broadcast station, the number of radios increased from eight million in 1965 to 75 million in 1976 (Zhao, 2004).

In the Great Leap Forward (from 1958–1960), it was very difficult for every household to buy a radio let alone television set, however, the loudspeaker fit the situation. Amplifiers with radios were widely used in most parts of China, becoming popular because of low cost and one radio being able to serve many people. For instance, many villages installed a large horn called *dalaba* (大喇叭); with an amplifier, the information transferred by the radio could be heard by the whole village. They broadcast news in the morning, lunchtime and evening, very loudly, which aimed to inform news to everybody. With the loudspeaker, every individual in the village could be guaranteed to receive news and information at the same time.

For people listening without pictures, it was hard to judge the reality and the information that was heard. Having to concentrate on the content of the broadcast and with only limited information presented without pictures made it difficult to think deeply, therefore at the beginning, the government preferred radio over television to spread their ideology. The spread of radios helped the central government stabilize the situation; the whole nation experienced extreme solidarity, which made an obvious contrast to the bad economic performance and the difficult international situation.[28]

Humans are curious and this natural ability pushed some audiences to try to hear more voices from the outside, but only very cautiously, since they might face punishment if caught. In 2015, I did an investigation in Dangshan, a town which belongs to Anhui province. While investigating elderly people who had experienced the time of Mao's rule, one man said *"the more government controlled the mass media, the more individuals wanted to hear other voices, especially when the government was at risk. The American voice, at the beginning when the new government was established, broadcast all content in Chinese. Most of the content was about the dictatorship of the government and worship of Mao, however, the Chinese media content was very different, being more focused on the improvement and advantages of the new government. We listened to the 'enemy' radio even though it was illegal, so sometimes it was risky to listen to the radio from outside and the quality of the sound was bad, it was hard to hear the content*

---

28  The People's Republic of China was not recognized by the West before the 1970s.

*clearly."* (Xu, male, high school, 67, Dangshan, personal interview, December 3, 2013, 16)

However, even though some people were aware of different voices, their opinions were not easily spread via the radio. These small groups of people tried to get to know the outside and wanted to hear different views of China from other countries, but this did not help to change the situation as the majority of people were getting information from the loudspeaker. According to group psychology, when the majority receive the same information, the minority who get different information will not be able to shake the majority opinions (Kohut, 1976). At the end of the Cultural Revolution, the central government started to decrease control of the media. The radio programs expanded to local government and private companies that could make their own programs (Ai, 2002). This strict control situation was also changed when radio was replaced by the television. With television, the government permitted foreign culture broadcasts and introduced foreign countries to the masses as the new government gradually established diplomatic relations with other countries, even with the so-called "imperialist countries" in Europe and America.

## Television

On March 18th in 1958, China produced their first black and white television set. When the television sets started to appear on the market, they were regarded as a symbol of the 'rich'. The high price of the television set corresponded to two years' income of an ordinary family, therefore, only rich people could afford it. At the early stage, before and during the Cultural Revolution, it was estimated that only 20 % of households in China had television sets (Television, 2012).

In 1958, the Beijing Television Station was established as planned. The leaders of the government required China to establish television stations as soon as possible. At the same time, black-white and color televisions were required to develop in parallel. This urgent mission was pushed by circumstance due to the Socialist Camp's conflict with the West. This sensitive atmosphere forced the Chinese government to make decisions to develop their own color televisions and broadcast stations as soon as possible in order to avoid "cultural colonialism" of the Western countries.

In the Socialist Camp, the east European countries and the Soviet Union had already established their own television broadcast stations, the Chinese government was worried about falling behind, and as a result, many broadcast stations were established in China even as the whole country was faced with poverty and found itself in a very difficult situation. The Beijing Television Broadcast Station was supposed to invest 95,000 dollars to set up broadcast stations. Because of lack of money, it only invested 17,000 dollars and established a station within five months. Besides the Beijing Television Broadcast Station, many television stations were established within a short term. However, compared to the rocket-like development of broadcast stations, there was not enough audience and a lack of programs (Zhao, 2004). In order to change the laggard situation, the central government made it very clear that China should diffuse the television as fast as possible.

The television diffusion during Mao's period can be divided into two stages. The first stage was from 1958 to 1966. During the Great Leap Forward period, the television stations increased to 29 and surged to 36 within three years. The Beijing Television Station[29] broadcast programs four times a week for about two to three hours per week. In the beginning, the programs were very simple because of lack of broadcast materials. Movies and comedy were the main programs, which accounted for 90 % of the programs, and comprised little news. The audiences were very homogenous, since most audiences were people who worked for the government. Ordinary people had to buy tickets to watch television programs. After the Great Leap Forward failed and was followed by the economic recession, the number of television programs decreased dramatically. In the end, the government decided to keep only eight television stations just for necessary work (Ai, 2002). Given the high price, little demand, and rigid marketing policy, television had grown sluggishly.

The second stage was from 1967 to 1976. On February 6th in 1967, the Beijing Television Station that was the symbol of the central government broadcasting was also forced to close. After 1968, as the domestic situation got better, the closed television stations were gradually opened again. The central government required every province, autonomous region, and independently administered municipal district to establish their own television stations. As a result, there were about 80 television stations by the end of 1971 (Television, 2016). Since the late 1970s, the penetration of the television started to grow slightly. According to statistics from the Ministry of Commerce and Central Television Station, by the end of 1975, there were 463 thousand television sets in China and 68 % of television households lived in cities; there were about 6000 color television sets (Television, 2012; TV, 2012).

The Great Leap Forward and the Cultural Revolution had dramatically jeopardized the development of television. During the Cultural Revolution, the

---

29 Former name was China Central Television.

government strictly controlled television content. As a result, the penetration rate of television plummeted to the lowest point, yet, the radio spread extremely fast. During the Cultural Revolution, programs that related to history, heroes, war, enemies, workers, farmers, humanity, ethics, morals, and love, etc., were all supervised by the government. Because of the strict censorship system, the broadcast stations had no material to present. As a result, except for the Shanghai Television Station, the other television stations had to close temporarily (Wedam, 1989).

The television content lagged behind the fast development of television stations. There were three video formats in the world at that time, namely PAL, NTSC, and SECAM. The formats were incompatible with each other. Considering the special international situation that China was isolated economically and militarily from Western countries, the Chinese government decided to develop their own video formats (History, 2012; Television, 2016).

The video formats in China were different from other countries so that the color television broadcast hardly made any progress in the beginning. After several years' development, the hope of the Chinese government to design its own format finally failed. In 1972, America's President Richard Milhous Nixon visited China, which changed the relationships among China, America and Western countries. After many negotiations and careful comparisons, China decided to choose PAL—the technology from the Federal Republic of Germany. Since China adopted PAL as a format, the color television started to diffuse. On May 1st in 1977, the Beijing Television Station first broadcast color television programs, and was then followed by Shanghai Television Station, Tianjin Television Station and Chengdu Television Station. As the development of color television thrived, the contents of televisions were filled with fresh values and vivid images. Moreover, many foreign movies and TV dramas were imported, which helped audiences to get familiar with the outside world (Television, 2012; Television, 2016).

## Fixed-phones

The first local fixed-phone was born in Nanjing in 1900. Before the People's Republic of China was established, post and telecommunications were controlled by foreigners. The diffusion of telephones was relatively slow. By 1949 the diffusion rate of fixed-phones was only 0.06 %, and there were about 218 thousand fixed-phone subscribers (Zhang & Zheng, 2011).

After 1949, the central government set up the Ministry of Posts and Telecommunications in charge of telephone services and the telecommunications industry was controlled under the planned economy. To a certain extent, during this time the real sense of telecommunications marketing did not exist since the telecommunications industry served as public goods, which were controlled by the government. Between 1949 and 1979, if ordinary people wanted to use a telephone, they had to go to the Bureau of Posts and Telecommunications (History, 2011).

The press in Mao's period was not independent, had a hierarchy system and was largely dependent on the government policy. Rather than working as a watchdog to supervise the government, it was an organ of the central government. Compared to television, radios played a more important role in China for society maladjustment and the government preferred to use radio to control society. The old media could easily be controlled by the government compared to the new media. In Mao's time, the traditional media controlled the information and ICT, and technology such as fixed-phones, was just at an early stage of diffusion.

## The Post-Mao Period: Learning Experience and Fast Diffusion of ICT

An investigation in 2013 in Anhui province suggests that in a social disorder period, government leaders will be more secretive compared to a period of social stability time. During the investigation, an interesting topic had arisen—changes of distance between a leader and the masses. In the semi-structured interviews, by asking questions such as how much information the masses could get from the mass media and how much they knew about their leaders, it was suggested that in Mao's period, leaders liked to expose their private life and the masses knew very well about their leaders' family and their daily lives. After Mao's period, from 1975 to 2009, leaders have become increasingly secretive to the masses. This privacy means that the distance between the masses and leaders has increased.

Leaders keep a certain distance in socially stable periods compared to times of disorder, which means they are more secretive compared to a period of social stability. The political usage of media is differential in Eastern and Western countries. The media are used for campaigns in Western countries in the name of political engagement and play a unique role in transmitting information, spreading values of candidates and supplying most of the information for people to use in voting. By providing access for politicians to convey their campaign promises to the electorate, the political concern is changed into a media news bias, and thus translates into a policy bias that large groups receive favorable policies, allowing the media to affect public policy design. This is the main political function of media, which is also the most obvious form that can be easily observed in the West.

Marx said "the most obvious advantage of the press is because it can intervene in social movement, becoming the mouthpiece of social movements. Besides, it can also reflect the whole situation and interact with the masses" (Liu & Chen, 1988a). Lenin holds that the press is not only the propagandist and agitator, but that it serves more like an organizer (Xia, 2006). As the biggest communist country, in China politicians are seldom seen having a debate and it is rare to see them make promises in front of the TV or in the press. The politician in China is more secretive compared to other countries and tends to keep a certain distance from the masses, the distance depending on the time period and social conditions. The media in China is more like a governmental organ instead of a Fourth Estate. The activist leaders, such as Mao, Lenin, Saul Alinsky, and Martin Luther King, tended to use social resources to enunciate general principles as the tactics and strategy to mobilize the masses and in the end overcome enemies and a hostile environment (McCarthy & Zald, 1977).

When China was established, the media exposed many stories and personal life details about Mao and other leaders such as Zhou Enlai, Liu Shaoqi, Zhu De, etc. The masses enjoyed this kind of news and stories and as one said, *"when we hear their stories, we feel near to the leaders, and we feel like we are living in the same family. We know how many children Mao has, how many wives he had, and we know our premier Zhou Enlai did not have children. We know how leaders and their wives got married, and we even know many cigarettes they smoke, which brands they like."* (Xu, male, high school, 67, Dangshan, personal interview, December 8, 2013, 16)

In 1975, Deng Xiaoping became the leader of China, and during his time the distance between leaders and the masses changed. The media mainly focused on what kinds of work and effort the leaders did for the country and most of this news came from the official media. The leaders kept a distance from the masses, therefore the meaning of "leader" included authority, distance, respect, fear and so on, whereas "leader" in Mao's period meant kind and familiar.

This change is like a coin with two sides. On the one hand, the leader and central government are endowed with more authority and deterrent force. As China already had a stable environment both inside and outside, and the reform and opening needed a strong government, therefore, keeping a certain distance between leaders and the masses was needed by the society; on the other hand, the increased distance might result in a bad relationship between leaders and the masses, making the society unstable and distrustful. One example is that since the middle of 1990, the rate of petitioners is increasing so that the government has to spend significant capital to keep the society stable.

Since 2013, the Seventh National People's Congress elected the new leadership of China. Hence, the distance between leaders and masses has changed again, given that the new leaders like to show their family life by designing a cartoon story, which aims to make the masses close with the new leadership group. Leaders such as Xi Jinping and Li Keqiang expose considerable information about their family and their youth life via mass media, and the official mass media has published many programs in order to introduce the new leadership's private life, not just their work career.

The distance between leaders and the masses can be regarded as an important indicator to measure the social condition. For example, it suggests that during the modernization process, the leadership used ICT to adjust the goals of society and social members' activity. In Mao's period, because of the special situation, the leaders successfully mobilized the masses by traditional media, such as radio. The central government mobilizing the masses invokes the Great Leap Forward, the Cultural Revolution and the insane enthusiasm of the Red Guards.[30] It is worth noting that the role of ICT during these social movements was a crucial tool for lessening the distance between leaders and the masses. Similarly, a large distance could lead to rigidity, which might jeopardize the cohesion of the society.

In 1978, Deng Xiaoping became the leader. After he witnessed the power of the masses and saw the masses lose control with the resulting social chaos, he said, China should not repeat the mobilization of the masses road. He initiated the economic reform process, which transformed China into one of the world's largest economies. In Deng's leadership group, the media were less focused on the moral ethical issues and class struggle (which is the important part of mass media in Mao's period), and more focused on economics. As Deng believed "No matter if the cat is black or white, if it catches mice, it is a good cat", and the value of "let some people get rich first" which suggest his strong will to develop the economy and less of a focus on social mobilization.

---

30 Red Guard has two meanings, the narrow meaning only means the student, the broad meaning points to all people who have pledged their loyalty to Mao Zedong, including workers, farmers and so on. In general, the Red Guard was mobilized by Mao Zedong during the Cultural Revolution, their mission was against the people who had different opinions than Mao, Capitalism, Confucianism, etc. However, their movement gradually lost control and changed into a historical disaster as many famous temples, shrines, books and other heritage sites in China were destroyed.

The process of developing countries importing advanced technology from the developed countries is called the advantage of the laggards. It reduces research costs and the diffusion time of technologies for the developing countries are short (Salomon & Jin, 2008; Wei, 2008). For instance, it took less than a decade to upgrade black-white television into color television and fuel the domestic television industry as well as stimulate a competitive television market.

## Press

After 1978, the central government started to reflect on the Cultural Revolution and gradually granted the press more freedom. The article *Practice is the Sole Criterion for Testing Truth*, which was published in Guangming Daily (*Guangming ribao* 光明日报), ignited a debate on the standards of the truth. As a result, reform and opening policy started and the newspaper turned back to its normal routine— the dark time of the press was ended (Shen, 1997).

The press changed from state-owned into semi-commercial, semi-state owned. The hierarchical press system has changed many times since China was established in 1949, but the essence of the press has never changed. A strict press hierarchical system limits the rights of the newspapers that low-level newspaper agencies cannot exceed the level to supervise high-level administrations or newspaper agencies.

Between 1950 and 2000, the number of newspapers increased nearly ten times compared to the number before. Some newspaper groups merged into a big media empire, such as Xinhua News Agency (*Xinhua she* 新华社), Nanfang Daily (*Nanfang ribao* 南方日报), Yangcheng Evening Newspaper Group (*Yangcheng wanbao* 羊城晚报) and so forth. Most of the newspaper subscribers are local government offices, institutions, schools and companies. These are the pillars that support the newspaper industry.

The press in China used to be distributed to each level of local government. The local governments and organs had the obligation to subscribe to newspapers. With such a distribution system, the newspapers had a sufficient number of readers. Since 1999, individuals who subscribe to newspapers are decreasing rapidly, especially with digital devices prevalent among the youth; most subscribers of the newspapers are elderly, literate, city citizens.

After 1990s the central government began loosening their control of the press. Some of the local newspaper offices have been withdrawn and reorganized, others merged into several big press groups. This change forms the three biggest press zones, namely the Beijing, Nanjing, and Guangzhou press zones (see Fig. 3-5).

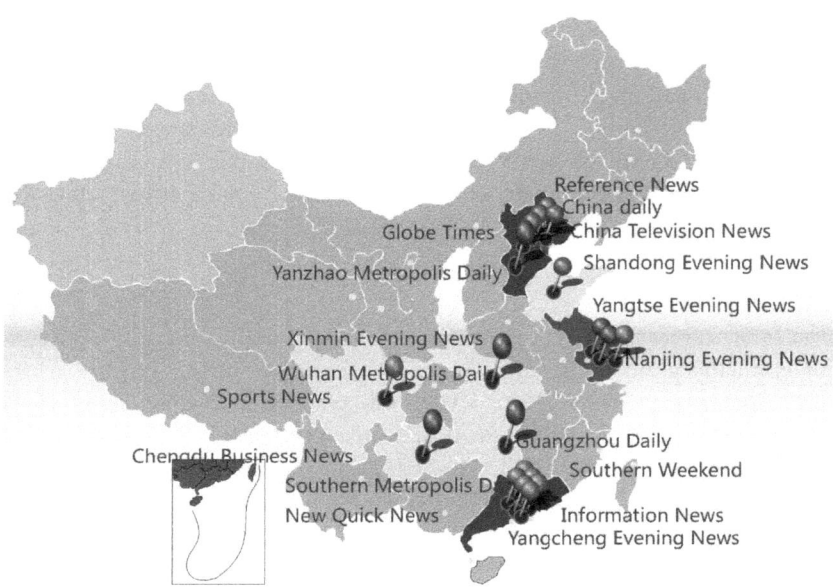

**Fig. 3-5:** Large newspaper groups in China
Source: Author's calculation.
Note: The map is designed according to the location of the newspaper agencies and circulation of newspapers, data collected from International Federation of Audit Bureaux of Circulations and World Association of Newspapers and News Publishers.

Fig. 3-5 shows the location of all the large newspaper publishers with a circulation of more than 100 million copies. According to the map, almost all of the main newspaper publishers are located in the middle and east part of China. Beijing, Shanghai and Guangzhou are the three important centers of the press. To some extent, Beijing represents the north part of China, Shanghai is the center of the Middle East part of China and Guangzhou is the agency of the south part of China. Besides geographical differences of the press centers, the contents of the centers are different as well. The press center of Beijing is more politically oriented, which focuses on political news and moral education, whereas Guangzhou is more critical about contemporary issues and pursues the liberty of the press. Shanghai is more neutral compared to Beijing and Guangzhou.

The Southern Weekend incident in 2013 can be seen as the voice of the press from the southern region asking for more freedom. The New Year's special theme called "Chinese Dream" aroused a conflict between the Propaganda

Department of Guangdong Province and the editorial department of Southern Weekend.[31] Before the articles were published, they were checked many times by the Guangdong government and in the end were replaced by another article, which was written by the Propaganda Department of Guangdong Province. This incident caused the newsroom staff of Southern Weekend to go on strike to fight for freedom of speech and against the censorship system. The strike quickly spread to the masses and many citizens supported the Southern Weekend. They demonstrated outside the gates of the Southern Weekend (Nanfang zhoumo, 南方周末) newspaper and the Propaganda Department of Guangdong Province in Guangzhou. After the strike, the central government became more sensitive about the press, especially the press in the southern region. However, why does the local government have authority to supervise local newspapers? In order to answer this question, I will introduce the hierarchy press system of China:

Newspaper agency A is a party organ of B town, hence, it has the same administrative level as B town, but it cannot supervise provinces or cities with administrative levels that are higher than A. A can supervise other towns and villages at the same level or lower than its own level. For example, if C town is at the same administrative level as B town, then A can supervise it; if the administrative level of D town is lower than A, A can also supervise D.

The newspaper agencies prefer to supervise other local governments or press which are at the same level or lower level, and pay less attention to supervise the B town or the villages controlled by the B town, because A is set up by B town. Theoretically, the authority of A is parallel to B town, but in reality A, is under control of the B town, which means A's exposure of fraud or corruption of another town is much easier to be passed by the government of B town. In China, a village is a basic administration, hence, all villages are supposed to be supervised by A. This can be used to explain why in China the news about corruption and fraud mostly happens in towns and villages, and illustrates the reason of conflict between Southern Weekend and the Propaganda Department of Guangdong Province. As they are at the same authority level, theoretically, the Southern Weekend can supervise the Guangdong government and have their own opinions. However, it was set up by the local government and the local government wanted to control it. Certainly, the main reason for the conflict can be seen as a fight for the freedom of the press, but the hierarchy press system illustrates the logic behind the conflict. The economic reform created a wave in

---

31 A member of the Southern Newspaper Media Group.

the press field, setting off big changes in the newspaper market. The press is more commercial and the environment of the press is more competitive. The press not only exists as an administrative unit but also as a profit-making organization, which has to remain financially independent and cater to the public and government interests.

The dilemma situation of the old media is that they are controlled by the government but have to take responsibility for their own profits and losses. On one hand, the contents will be checked by the organs. On the other hand, in order to cater to readers, the press tries to produce sensational news to attract readers. The press has to stand carefully on the edge of legitimacy and try to find a balance in the muddy water. In the information age, the annual issues of newspapers have decreased worldwide. The same has happened in China as well. The big newspaper publishers work together to resist the influence of the Internet. However, compared to the new media, the traditional media can hardly compete with the new media. In the information age, the resource of information is not controlled by the government but can be generated by the masses. Media users get information-not passively, they have many accesses to get information and can spread them fast.

In summary, big newspaper agencies are party organs and cooperate with the government. In the south part of China, the newspaper is less politically oriented. The regional differences shape the various styles of the newspapers.

## Television

The fast television development was from 1978 to 1990. It was ignited by the reform and opening-up policy (*gaige kaifang* 改革开放). The central government set a target of modernization—China should focus on improving the light industry, agricultural sectors and tertiary industry. The government encouraged the media to support and serve the modernization. The television, which is an important technology driven by modernization, had been specifically called upon to play a crucial role in achieving these goals (Epstein, 1982). In 1979, the number of television sets surged to 4.85 million, three years later, there were 27.17 million television sets in China. By 1985, the Chinese television production exceeded the American; China became the second largest television producer after Japan. Two years later, China became the biggest color TV production country in the world (2012c, 2012d). Between 1978 and 1990, television sets were prevalent and widely adopted and were regarded as a symbol of the success of modernization by the central government (Lull, 2013). At the same time, in 1979 the television rate per hundred households between urban areas and villages was 17.2 and 0.8 respectively, which means the physical divide existed at the beginning stage.

With the wide penetration of television, from 1985 to 1993 the government policy helped to stimulate the television industry and decreased the gap. Many local brands, such as Panda, Venus, Peony and so forth emerged. The good price attracted many customers, which helped upgrade the black and white televisions into color televisions within a short time (2012d). However, the shortage of television programs invoked by the film publishing system became an obstacle compared to 'over-growing' television sets. In viewing the entire television developmental history, it is always intertwined with the ups and downs of the film industry (Wu, 1997). Since 1979, the film industry started to become financially independent, and it had to take responsibility for profits and losses, yet, prosperous television programs jeopardized the market of the film industry so that cinemas were quickly vanishing in China. With financial pressures, film companies stopped supplying films to television stations, which resulted in the television programs being largely reduced (Ai, 2002; Wu, 1997).

The conflict between television stations and film companies lasted for many years. In order to solve the problem, the Central Broadcasting Administration held a conference in an attempt to fix the problems between television programs and the film industry. The Central Broadcasting Administration asked television stations to develop their own programs, which meant the broadcast material should not be limited to films. The new requirement got rapid results, many programs designed by television stations became very popular, for example, Big Windmill (*da fengche* 大风车) for children, Half Sky (*banbian tian* 半边天) for women, and Red Sunset (*xiyang hong* 夕阳红) for the elderly, etc. The CCTV News and Topics in Focus (*jiaodian fangtan* 焦点访谈) are very successful examples that amassed a large audience and are still a very important part of television programs today (Zhao, 2004). As the television stations were allowed to produce their own programs, advertisements became an important source of profit. Shenrong Wine, RADO, and Citizen were the first companies to publish advertisements at Shanghai Television Station (History, 2012). The considerable revenue from advertisements provided financial support for broadcast stations to produce different programs.

In contrast, movie companies lost their customers as televisions proliferated throughout the whole country. Theaters and cinemas in China experienced the worst situation between 1980s and 1990s. Only some big cities had theaters and cinemas. Most of the towns and villages pulled down cinemas and replaced them with shopping malls. After a nearly two decades' decline, the film industry decided to depend on television platforms to broadcast movies.

The television media in China thrived from importing, imitating, and producing their own television programs. The flourishing of television programs applied much stress to the government, as many imported television dramas were not fit for the ideology of the government. The national broadcast station decided to encourage broadcast stations to design their own television dramas and not just import programs from other countries. In addition, television stations attempted to strike a balance between entertainment programs and news programs, which meant the broadcast stations had to highlight current issues and public discussions.

By 1991, the revenue from the advertisement on television reached one billion dollars, which was the first time television exceeded newspaper and radio, thus becoming the most popular medium for advertising. In 1992, the "South Tour Speeches" (*nanxun jianghua*南巡讲话) by Deng Xiaoping set off a new upsurge of the market economy. The central Television Station established their Marketing and Development Department—television stations started to cooperate with companies and the growth rate of advertising profit on television reached 105.4 % (History, 2012).

It is said that every coin has two sides. Behind the rosy scenario, the rigid, planned economic system had serious consequences. When faced with the transition period, the Chinese market largely depended on the government's adjustment, since the central government gave authority to local governments to establish their own programs for the sake of increasing viewership. By 1994, there were 3125 channels in China. This figure was more than all the channels of America (2606), Russia, Japan, France, Germany, UK, Canada, India, Australia, Brazil, and Pakistan combined. The central government realized there were too many television channels, so it quickly adjusted the policy to restrain the emerging programs and the number of channels decreased to 2587 by 2011, however, it was still too much. The diffusion of television and reform of television programs were closely related to the political climate (Ai, 2002). From the top to the grassroots, all television workers were appointed by the government, and the stations were owned by the government. Except for the government, neither state enterprise nor private company was allowed to invest in television stations. Comparable to the press industry, the television stations have to keep financially independent and also have to serve the country.

The government-oriented policy shows less efficiency in a rigid economic system; many television companies are state-owned and struggling to transition to the global economy. Chang Hong, one of the biggest television companies in China, started to reduce its price since the Chinese government is committed to reducing television costs, which aims to ensure that every household in

China will have television. Afterwards, a fierce price war ensued between television companies and many suffered huge losses. As noted earlier, newspaper publishers incorporated into big groups for better development. Similarly, television stations began merging, some forming huge conglomerates (History, 2012).

Between the 1980s and the 1990s, the Chinese government made great progress in the diffusion of television. The television that was regarded as a luxury appliance now became the most common medium in China. Digital Light Processing (DLP), Liquid Crystal Display (LCD), and Plasma Display Panels (PDP) gradually took the place of the tube television, demonstrating the arrival of a new age of television. During the Beijing Olympic Games (2008), television companies made advertisements such as: *buy a new television, watch Olympic Games*. The Chinese government regarded distribution of television as one of the important symbols of a successful modernization, which has reached its goal.

## The Learning Experience and Telecommunication Reforms

In the context of globalization, the developed countries are like precursors to spread new technologies. The laggard countries learn media policies from frontier countries based on the assumption that the media somehow have similar natures, which will result in similar outcomes (Grin & Loeber, 2006; Simmons & Elkins, 2004; Stone, 2001). With regard to technology transfer, the developing countries may easily hitch a ride on innovation with less investment of new technology, but can nevertheless benefit from mature, advanced techniques to accelerate the penetration procedure, which is called "the advantages of the backward" (Glaziev, 1991). Mostly, the policy transfer is distinguished by the policy subsystem, which can hardly be defined without a certain historical-institutional analysis (Sabatier, 1986). Therefore, to understand one country's telecommunication industry and polices, it would be important to reflect and explore about whether and how catch-up occurs.

The policy learning studies were burgeoning from 1980 to the late 1990s. The notion of policy transfer is "a process in which knowledge about policies, administrative arrangements, institution etc. in one time and/or place is used in the development of policies, administrative arrangements and institutions in another time and/or place" (D. Dolowitz & Marsh, 1996, p. 343). The study of policy learning can be traced back to Bennet and Howlett's 1992 article: *The Lessons of Learning: Reconciling Theories of Policy Learning and Policy Change*. From that time onwards, the learning theories have experienced several major changes (C. J. Bennett & Howlett, 1992; Meseguer, 2005). The range of learning was broadened from governmental actors to societal actors; the study focus was no longer confined to a single domain, but the learning process became viewed as a collective act more than an individual act. The big step of policy analysis is not only to specify policy learning, but also broaden the relationship between agency, structured learning, societal change, the Third Sector and so on (Grin & Loeber, 2006).

The most important question when policy transfer from one domain to another takes place is: who are the learning actors? The bureaucrats, politicians and state agencies are the official actors in the processes of transfer policies. As countries interact with each other more frequently than before, other forces such as transnational networks and non-state actors (non-governmental organizations or the Third Sector in Governance), and the knowledge organizations, for instance, universities, scholars, foundations, consultancy, and think tanks also play a crucial role in transfer policies and as well as in policy formulation (Stone, 1999, 2001).

Hall outlines three important factors in the learning procedure, namely referencing policies, intellectuals or knowledge organizations, and the capacity of states. He argues that "one of the principal factors affecting policy at time-1 is policy at time-0" (Hall, 1993, p. 277), the time lag between domains of policy make policy learning available. Social learning, as defined by Hall, is "a deliberate attempt to adjust the goals or techniques of policy in response to past experience and new information" (Hall, 1993, p. 278). Second, the effort of intellectuals, experts and knowledge organizations are three important elements who push forward the learning process at the "interface between the bureaucracy and the intellectual enclaves of society" (Hall, 1993, p. 277). Finally, yet importantly, based on the domestic context, whether the state has the capability to apply the policy determines its success and failure.

With a country as an incubator, when it adopts technologies, especially communication technologies, the government always learns from the frontier countries' experience in order to set up their policies and implement laws. Richard Rose pointed out that policy-makers could benefit from learning and refereeing policies from other domains when they deal with their own issues in terms of comparable problems (D. Dolowitz & Marsh, 1996; Rose, 1991; Stone, 1999). The frontier countries succeeded with a set of policies and received positive results, the laggard countries followed, drawing from the great successes of the Asian Tigers as well as countries like China and some Latin American countries (Meseguer, 2005). The economic burst of Asian countries and the development of South American countries have attracted more eyes to policy transfer in the area (D. Dolowitz & Marsh, 1996; Lingard, 2010).

One consequence of the policy learning process is policy failure. D. Dolowitz and Marsh (1996), based on the work of Bennett and Rose, have developed three types of policy transfer and further explored reasons for failure. They regard transfer as a variable, which can be classified as below:

Voluntary transfer—domestic situation such as the society's dissatisfaction with the social-economic situation, society structure or current policy that will motivate the voluntary transfer.

Direct coercive transfer—the Third Sector such as international corporate (WTO, OECD, etc.) or NGO consultants blurs the boundary and marginalizes economic entities, countries who engage in these Third Sectors have the obligation to adjust policies to cooperate with them.

Indirect coercive transfer—different from the last two policy transfer types, the indirect coercive transfer is more an initiative and positive change than a passive change. The external interdependence links the world tighter than before, the giant multinational corporations combine their profits from many countries, not only one country.

Further, they mentioned three factors that contribute to explaining policy failure namely uninformed transfer, incomplete transfer and inappropriate transfer. Either the failure of policy transfer happens because of lack of sufficient information or the learned policy cannot fit the context, or both. Here, the factors Dolowitz and Marsh have mentioned are similar to the basic factors in Hall's policy learning procedure. The policy information, learning actors and context are the crucial elements.

A policy set-up procedure includes policy-makers, decision makers, operational staff and target groups, as well as the policy itself and its subsystems. "Learning occurs when policy-makers adjust their cognitive understanding of policy development and modify policy in the light of knowledge gained from past policy experience. The idea of policy learning is informed by an understanding of policy failure providing impetus to place new ideas on the policy and political agendas. With increasing policy failures, greater interest is shown in alternative ideas and politicians will have particularly strong incentives to seek out and embrace ideas that challenge the policies of their opponents" (Hall, 1993, p. 73).

At the beginning, the telephone as the communication medium was under the cloak of politics in China. From the three decades from the 1950s to the 1970s, the central government invested one billion dollars into the fixed-phone, which was far from supporting the telecommunications development at the threshold. After 30 years' development, the average growth rate of fixed-phones started to be improved but was only 7.8 % per year. In 1978, the number of fixed-phone subscribers was still only 1.9 million and tele-density was 0.38 %, which was lower than one-tenth of the world average. The population of China accounted for one-fifth of the whole world population at that time, yet the percentage of fixed-phones occupied less than 1 % of the world. Even though in some major cities, for example, Beijing, the fixed-phone was already available in the city, 20 % of long-distance calls could not connect and 15 % of the calls had to wait for one hour (Zheng & Zhang, 2011). Some films describing that age often show people bringing food along with them to wait in line for the telephone.

During the reform and open up, the slow pace of telecommunications became an obstacle of economic development. The government gave priority to developing telecommunications by taking several measures. Firstly, huge amounts of money were invested in the telecommunications industry from different capital sources including central and local governments, collective and individual investments, but the foreign investment was not permitted for telecommunications development at that time. Secondly, taxation for the telecommunications industry was reduced with the aim of guaranteeing enough money for technological innovation. The genesis of the fixed-phone installation and usage fees were relatively expensive along with burgeoning additional costs. However, these high costs helped the telecommunications industry to accumulate capital. From 1980 to 1989, the fixed assets investment in the telecommunications industry reached 15 billion dollars. As noted earlier, compared to the one billion dollars invested between 1940 and 1979, this intensive funding was regarded as the lifeblood and sinews of the telecommunications industry (Cai, 2003; Gao, 2008).

Telecommunication reform in China was from 1992 to 2008, and was led by the central Chinese government. As people's living standard improved, the telephone became an important household appliance. However, the old monopoly system of telecommunications could not satisfy the market. The government realized it was necessary to allow other companies or industries to enter into the telecommunications market and improve the vitality of the market. In 1993, the State Council decided to commit a reform in telecommunications. The government distributed assignments for the big telecommunications companies in charge of different areas such as the phone sector, railway telecommunications sector, network sector, and satellite communication sector, respectively. As a result, the four big telecommunications companies divided the "cake"; yet, all of them were directly controlled by the government.

In 1992, the Electronics department, the Power Department, and the Railway department proposed to establish China Unicom Corporation, given that the Ministry of Posts and Telecommunications cornered the market which jeopardized the development of telecommunications. The Ministry of Posts and Telecommunication, of course, was strongly against this proposal. Their claim referred to other countries' experience where the telecommunications industry was all monopoly at the threshold (Zhou, 2001).

The experience of the UK's telecommunications reform inspired the State Council to pass this proposal. As a result, China Unicom Corporation was set up at the end of 1994 (Gao, 2008). China Unicom Corporation chose to develop the sector with the most potential: mobile phone service, as their main service, since mobile phone infrastructure needed less investment but was endowed with high profits, which made it easier to develop at the beginning. Afterwards, the Ministry of Posts and Telecommunications could not play the role of both administrators and operators any longer, and it had to transform its role to administrative management. Some functions were separated by setting up a new cooperation—China Telecom Corporation. The telecommunication reform of China had started.

Breaking up monopoly companies is difficult in view of some cases from developed countries. When China decided to perform a telecommunications reform, it faced many obstacles. The central government, local governments and the telecommunications companies joined the debate about telecommunications separation, since China Telecom was the only company providing the services.

The road to establish a balanced telecommunicates system is hard. To get a better way to understand the policy learning process and how it affected the development of ICT in China, it is worth noting that the four telecommunications reforms that began in 1998 were finished in 2008.

## The First Reform

The strong economies of the developed countries convinced the developing countries to believe that the laws and regulatory systems from developed countries were effective. The developed countries as frontier countries took initiative to adopt the new technologies and provided a helpful reference for the developing countries. In order to commodify telecommunications technologies, the Chinese government referenced developed countries' experience and started the reform of telecommunications. In view of the experience of the UK, the new telecommunications company should occupy 10 %~15 % of the market within five years. The Chinese government referenced the UK telecommunications reform to set up the first reform. So, what is the result five years later?

The government-oriented policy made temporary progress. By 1994, the adopters were only hundreds of scientists. However, just five years later, in 2000, this figure reached 8.8 million. The China Internet Network Information Center (CNNIC), which was set up on June 3rd, 1997, aimed to study Internet development and report annually (Zhou, 2001). It helped the central government to design policy and arrange the next step of Internet development.

According to the goal of the government, China Unicom Corporation, which is the second biggest telecommunications company in China, should occupy 30 % of the market by the year 2000. However, as it turned out, the reform failed to achieve the expected target. After five years' effort, China Unicom only occupied 3 % of the market and more than 95 % fixed-phone market was occupied by China Telecom, which was the biggest telecommunications company in China. The assets of China Unicom was only 1/260 of the assets of China Telecom (Zheng & Zhang, 2011). The top company was still in the dominating role whereas laggard companies had less chance to gain market share. The first reform of the telecommunications industry failed to reach the expectations of the government.

In terms of China, the planned economy is one of the important reasons to attribute this failure. The telecommunications companies could not compete equally and both of them were government organs. The rigid, politically oriented market could not regulate price by itself, with a hierarchical telecommunications system, senior monopoly companies influenced market price by setting technical barriers and price barriers towards younger companies. Given these substantial inequalities, it was difficult for the peripheral and laggard companies to compete with the big companies within a short time.

The competition of duopoly accelerated the speed of telecommunications development. At the same time, on March 11th, 1998, the Ministry of Industry and Information Technology was established which is in charge of the telecommunications industry (Cai, 2003). As the Ministry of Posts and Telecommunication already existed to manage the telecommunications industry, why did the government set up another ministry in charge of telecommunications? As mentioned earlier, the Ministry of Posts and Telecommunications had separated part of its functions to China Mobile Corporation and it served as an administrative agency at the same time. However, essentially it was still setting rules and running businesses, as China Telecom Corporation once belonged to the Ministry of Posts and Tele Communication. Thus, regulations and policies designed by the Ministry of Posts and Telecommunications were imposed upon China Telecom Corporation. The establishment of the new administration helped to coordinate different telecommunications corporations and indicated the second reform was under deliberation.

## The Second Reform

The second telecommunications reform was implemented in 2001 China Jitong Corporation, China Netcom Corporation, and China Tietong Corporation were set up in 1994, 1999 and 2001 respectively (Mu & Lee, 2005) China Telecom Corporation was still the biggest telecommunications company and telecommunications in China was still a monopoly industry. To unravel this issue, in February 1999, the State Council accepted a proposal to reunite telecommunications to establish a fair market environment. This reform intended to weaken China Telecom Corporation and to encourage more telecommunications companies to enter the market. As a result, China Telecom Corporation was separated into four parts—fixed-phone, mobile phone, pager and satellite communication. Every branch-corporation mainly focused on one telecommunications field. The old giant China Telecom Corporation was replaced by four companies: the new China Telecom Corporation was authorized for using the band and grids of the original one, managing fixed-phone service; China Mobile Corporation was in charge of mobile phone service; Guoxin Corporation ran the pager business and China Satcom Group operated satellite telecommunications (Cai, 2003; Gao, 2008; Zhou, 2001).

Since China Unicom Corporation did not have its own telephone grids, it had to rent grids from China Telecom Corporation, and its market share was below 5 % which meant it still could not compete with China Telecom Corporation. When China Telecom Corporation was separated into four parts, the Ministry of Industry and Information Technology endowed more rights and business to China Unicom Corporation. The steps included expanding the range of its business, allowing the price of China Unicom Corporation to be 10 %-20 % cheaper than China Mobile Corporation's, merging other companies into China Unicom Corporation, giving priority to China Unicom Corporation by investing money and offering the Code Division Multiple Access (CDMA) technology, etc.

After the reform, three big telecommunications corporations shared the fixed-phone market, breaking the monopoly of China Telecom Corporation. The above measures intended to balance the market and form a competitive system. However, as the political-oriented industry cannot freely adjust to the market, after the separation China Telecom Corporation still accounted for more than half of the revenue of the fixed-phone.

The good outcome was that the series of reforms interlocked with other consequences immediately, such as lower cost, better services, and advanced devices. The telecommunication industry was growing faster than the GDP, thus boosting the media industry not only in the hardware field such as cell phones and computers, but also in the field of E-commerce as well as entertainment games.

The first and second reforms indicate a progressive trend whereas a big monopoly might be divided into small parts for the sake of market equality. For instance, China Telecom Corporation was divided into four parts. After the reform, one of the outstanding effects was that China Unicom Corporation became the company that covered every spectrum of telecommunications services—fixed-phone, mobile phone, Internet Protocol call, and pager, which might challenge China Telecom Corporation's development. Yet, there was a striking difference between China Telecom Corporation and the other companies, even though it partitioned some of the services and established new companies, it still occupied the majority of the market in fixed-phone and Internet; 99.7 % of all local calls and 97 % of all distance calls were provided by China Telecom Corporation, which also dominated 97.8 % of the network service.

**The Third Reform**

Shortly after the second reform, on December 11th, 2001, the Ministry of Industry and Information Technology approved a plan to divide the new China Telecom Corporation into two parts, the south part and the north part, which represented the third telecommunications reform. There were 21 provinces, 70 % of the telecommunications services of which belonged to the south China Telecom Corporation, and the other part continued using the band of the China Telecom Corporation. The north part had ten provinces, which comprised 30 % of the telecommunications services. By combining with the China Netcom Corporation and the China Jitong Corporation, a new company was established—China Netcom Group Corporation Limited. Since the third reform, every telecommunications sector had two companies to compete with and no one could occupy more than 50 % of the market share (Shen, 2005).

The third reform made significant progress. In 2007, telecommunications revenue was two times that of 2001, increasing 11 % every year. The subscribers surged from 326 million to 913 million, among them 547 million subscribers who were mobile phone users (Cai, 2003; Zheng & Zhang, 2011). China rocketed to the biggest market in both the fixed-phone and mobile phone fields. During this time, mobile phones succeeded fixed-phones to become the most popular medium. Consequently, the old market pattern had to change to fit the new system.

Tab. 3-1: Authors' Calculations based on CNNIC Statistics in 2000 and 2002

| Year | Local call (fixed-phone) | Distance call (fixed-phone) | Mobile Phone | Internet Protocol call |
|---|---|---|---|---|
| 2000 | China Telecom Corporation China Unicom Corporation | China Telecom Corporation | China Mobile Corporation China Unicom Corporation | China Telecom Corporation China Unicom Corporation China Netcom Corporation China Jitong Corporation |
| 2002 | China Telecom Corporation China Unicom Corporation China Netcom Corporation China Tietong Corporation | China Telecom Corporation (include international call) China Netcom Corporation (include international call) China Unicom Corporation China Tietong Corporation | China Mobile Corporation China Unicom Corporation | China Telecom Corporation China Netcom Corporation China Unicom Corporation China Mobile Corporation |

Source: Author's calculation based on CNNIC Statistics.

## The Fourth Reform

In order to integrate resources and protect local telecommunications companies in the 2008 worldwide economic crisis, China started the fourth telecommunications reform. The central government began a series of measures. For example, China Telecom Corporation merged the CDMA business of the China Unicom Corporation and the basic telecommunications business of China Satcom Group; merging China Unicom Corporation and the China Netcom Corporation into one company; the China Tietong Corporation was taken over by the China Mobile Corporation and separated the railway communication system to the Ministry of Railways in 2009 (Shen, 2005). As a result, the former six companies' competition changed into three big companies' competition—China Telecom Corporation, China Mobile Corporation, and China Unicom Corporation, becoming the three biggest companies sharing the telecommunications market.

Based on the above, I designed a graphic to illustrate (Fig. 3-6) the four reforms, which were expected to establish an integrated network with interlink and intercommunication mechanisms and achieve a competitive telecommunications market. It might be true that the earlier three reforms have changed the monopoly pattern. However, considering the fourth reform, China Telecom

Corporation controls the majority of fixed-phones and Internet services, China Mobile Corporation occupies the mobile phone market, and China Unicom Corporation is in charge of the remaining market of fixed-phones and mobile phone services, suggesting that the telecommunications market is still dominated by several big companies.

On September 26th, 2013, Zhang Feng, the Chief Engineer of the Ministry of Industry and Information Technology of the People's Republic of China, announced at the Broadband China forum that China was encouraging private companies to enter into the telecommunications industry, which legitimized the private telecommunications companies. The four reforms helped China to diffuse ICT. According to the 37th report from CNNIC, at the end of 2015, China had 0.688 billion Internet users, accounting for 50.3 % of the whole population. The number of mobile phone users is still growing. By the end of December 2015, mobile Internet users were 0.62 billion and comparable to computer Internet users—more than 90 % of mobile phone users accessed the Internet via mobile phone (CNNIC, 2016).[32]

The road of China's telecommunication reform was not smooth but it's relative success. The Chinese government played dual roles in the telecommunications industry: administrator and investor. Essentially, telecommunications were born as the organ of the government, and grew up in an environment that lacked competitive mechanisms. By referring experiences from developed countries, China's telecommunication reform did not take long process until it finished the transformation. After the reforms, China establishes a competitive mechanism in the telecommunication market.

## The Fast Diffusion of ICT

According to Rogers (2010), "Diffusion is the process by which an innovation is communicated through certain channels over time among the members of a social system. It is a special type of communication, in that the messages are concerned with new ideas" (Rogers, 2010, p. 5).

### Fixed-phone

In the 1990s, the diffusion of telephones witnessed the quickest progress. The fixed-phone dominated the telephone market, but at the same time, the nascent Beep

---

32 See http://www.cnnic.net.cn/hlwfzyj/hlwxzbg/201601/P020160122469130059846.pdf. Accessed April 6, 2016.

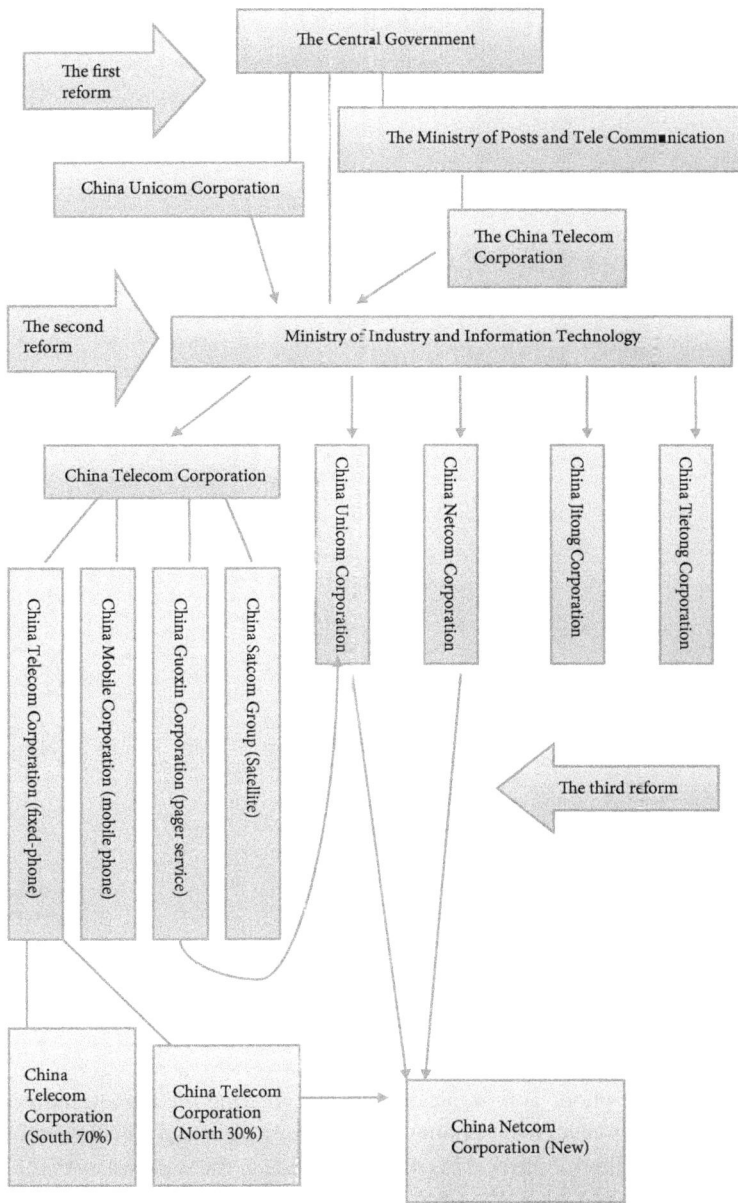

**Fig. 3-6:** The telecommunications reforms in China.
Source: Author's calculation.

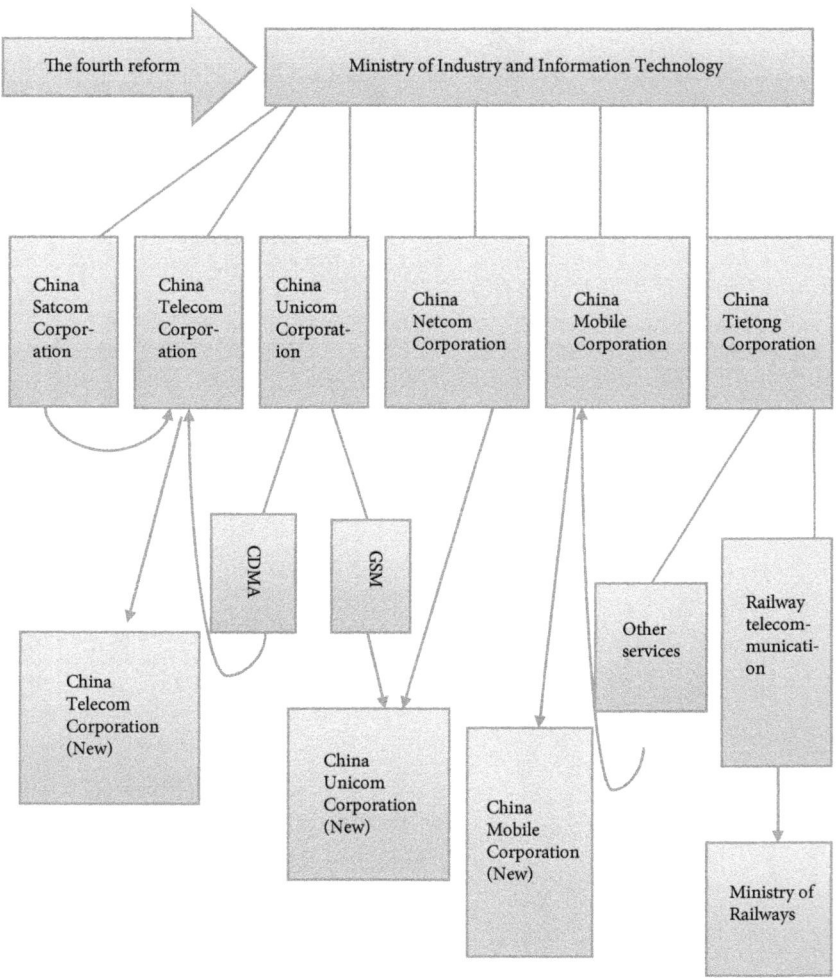

**Fig. 3-6:** Continued

Pager and mobile phone communicators started entering the telecommunication market as well. In 1990, fixed-phones could mainly be found in cities and there were only 5.2 million subscribers in cities. According to the statistics from the State Statistical Bureau of China, the fixed assets of the telecommunications industry entered into a high-speed growth period from 1992 to 1994, the growth rates were 99.0 %, 157.5 %, and 96.6 %, respectively (Zhang & Zheng, 2011). Previous policies had helped the telecommunications industry to accumulate capital boosting the telecommunications economy. Huge investments brought great success, since the infrastructure of telecommunications was already established, old equipment was updated, and the number of subscribers increased dramatically (Phone, 2012).

Since 2004 the expansion of fixed-phones slowed down, subscribers of fixed-phone peaked in 2006, and in 2007 the trend downward began. The contrast between the mobile phone and fixed-phone is sharp: in 1991, fixed-phone users were 176 times as many as mobile phone users, yet, by 2003 mobile phones exceeded fixed-phones for the first time and became the most popular communication devices in China (Zhang & Zheng, 2011; Zhang, 2006). The fixed-phone is gradually fading away in the media market and is being replaced by the mobile phone.

## Internet

Once the Internet arrived in China, it developed very fast and users benefited from the new information communication technology and also enjoyed the role of information publisher. However, the attitudes of the central government toward the information society are ambiguous and hesitant. The central government tries to open the door of information but they do not want to open it fully. Deng said China should "follow stones to cross the river", which means to achieve modernization, China should walk every step carefully. This doctrine has guided communist thought for so many years that the ruling party is very cautious to open the information age door. Since the Internet has become the most important part of ICT, it has changed the social order, influenced the way the government communicates with the masses, influenced inhabitants' daily life, and changed values of individuals

The spread of the Internet in China was influenced by certain individuals who played important roles in the social system and their opinion leadership[33] had a great effect on the diffusion procedure. The wildfire diffusion of the Internet in some developed countries started from the 1980s to 1990s and diffused to large scale between 1990 and 2000 (Crenshaw & Robison, 2006; Robison & Crenshaw, 2002; Alexander Van Deursen & Van Dijk, 2011). In the case of China, the Institute of Computer Application (ICA) in Beijing cooperated with the

---

33 Opinion leadership is a concept from the book Diffusion of Innovations from Everett M. Rogers. According to him, "Opinion leadership is the degree to which an individual is able to influence other individuals' attitudes or overt behavior informally in a desired way with relative frequency. It is a type of informal leadership, rather than a function of the individual's formal position or status in the system." Rogers, E. M. (2010). Diffusion of innovations: Simon and Schuster.

University of Karlsruhe to establish the Chinese Academic Network (CANET) to develop the Internet in 1986. On September 20, 1987, the first email, "Across the Great Wall we can reach every corner in the world", was successfully sent to the University of Karlsruhe via access through ITAPAC (Italy) and DATEX-P (Germany) and was regarded as the landmark case that opened the first page of the Internet in China (Internet, 2004).

However, in the same year, on the other side of the earth, the biggest Internet region had already set up a legal norm for the Internet called the *Computer Fraud and Abuse Act,* which was promulgated by the American government. At the same time, the Internet in the US started to be regulated by the spectrum of law, the Internet had just been introduced to China.

When the Internet was connected, the speed was very slow with 300bps being the standard. The email was available when the Tsinghua University adopted the X400 protocol from the University of British Columbia in Canada and connected with UBC via the X.25 network. By 1988, the Institute of High Energy Physics from the Chinese Academy of Sciences extended Digital Equipment Corporation net to China, which made cross-country network communication available (Internet, 2004). Following the trickle-down effect, the Internet first penetrated big cities and the elites, such as scientists and politicians in Beijing and Shanghai. Around one thousand scientists from different areas such as meteorology, geography, physics, astronomy, etc. were the first group of Internet users. They were the frontier Internet users and the backbone of Internet developers. At the beginning of Internet diffusion, Germany was an important back up which supported China's Internet development. All the email and documents were transmitted via Germany's access to the world.

In October 1989, a project called National Computing and Networking Facility (NCFC) was set up. It was sponsored by the World Bank and the central government, which aimed to establish physical access and to encourage software institutions to develop the Internet. Several years later, it became a famous "Silicon Valley" in China called Education and Scientific Research Demonstration Network, located at Zhongguancun District. On November 28th, 1990, Qian Tianbai, who was regarded as father of China's Internet, registered.CN as China's Top Domain Name System at Network Information Center of Stanford Research Institute, demonstrating China had its own identity. However, as the CN top-level domain server was temporarily set up at the University of Karlsruhe, China was only partly connected to the Internet (Internet, 2014; Xiang, 2009).

When China decided to join the world's Internet, many obstacles were set up by the frontier countries. In October 1991, Walter Toki, the chairman of American's delegation, suggested China join the Internet at the Sino-US High

Energy Physics conference. With hard negotiations on both sides, half the door of the Internet opened as the American government allowed China to connect their network to the Energy Department with additional requirements. For example, China could only connect to the Energy Sciences Network (ESnet) of America and to no other networks, every email needed to be checked by the US government, and China could not use the Internet in the military and business fields, etc. Only by accepting all these requirements, could China connect to the Internet via Energy Sciences Network (ESnet) of America (Internet, 2012).

At this time, the Chinese government realized the Internet will trigger the fourth industrial revolution and urgently wanted to join the Internet. One way to join the Internet was by seeking to get more support from other countries. In June 1992, the INET' 92 annual conference was held in Japan. A researcher named Qian Hualin, who came from the Chinese Academy of Sciences, met with the chairman of the International Network Department of the U.S. It was the first time that China officially brought up the issue of Internet connectivity, not only via the access of ESnet but to connect entirely. Unfortunately, the request was refused by the US government with the claim that China was a socialist country and the ideology was different from the US, besides, the US government assumed that the Internet contained vast information that related to the government and organs of US, and it also speculated that the free information and technological pictures might be utilized by the Chinese government to develop science and technology. Based on these reasons, the request of China to join the Internet was refused many times (Internet, 2004; Internet, 2012).

However, the difficult beginning did not dampen the will of joining the Internet, but rather stimulated more motivation by the Chinese government to develop the Internet. On March 12th, 1993, Vice Premier Zhu Rongji chaired the "Golden Bridge Project", a meeting proposed and deployed to set up the national network to provide economic information to the public. On August 27th, 1993, Premier Li Peng approved the budget of three million dollars to support the construction of the Golden Bridge Project. Meanwhile, by participating in Internet conferences and communicating with other countries, China was gradually accepted by Western countries. In 1993, China applied again to join the Internet. The delegation proposed to both the INET conference and Coordinating Committee for Intercontinental Research Networking conference and the conferences discussed whether China should join the Internet or not (Internet, 2004; Internet, 2012).

Participation at meetings and expanded mutual communication with other countries were rewarded with successful feedback. At the Sino-US Science and Technology Cooperation Conference, which was held in Washington in April

1994, the vice dean of the Chinese Academy of Sciences, Hu Qiheng, visited the National Science Foundation of US (NSF) to discuss the establishment of the first direct TCP/IP connection to China. She restated the Internet proposal that the Internet of China should connect to the NCFC to join the NSF of the United States. Finally, the proposal was approved allowing China to connect via the Internet to the world. This big event cheered up the whole country, which was regarded as the top ten technology news in 1994 and also listed as one of the most significant scientific technology achievements.

Ever since this historic time, the achievements in the Internet field have been remarkable. Following a carefully planned schedule, the basic function of the Internet, such as the use of email, was already ubiquitous among the frontier users. The next target was to expand Internet users. In March 1995, the Chinese Academy of Science took the first step to connect the Internet to big cities such as Shanghai, Nanjing, Wuhan, Hefei, etc. as the completion of the CHINANET network, and the nationwide spread came true.

The Internet was first applied to the science area and education field. As mentioned, the scientists in the fields of meteorology, geography, physics, and astronomy, as well as politicians in Beijing and Shanghai were the first Internet users. In 1996, the first Internet café appeared in Beijing, which provided access for users without personal physical access (Internet, 2004). The Internet café at the early diffusion stage played a significant role, as computers were still an expensive medium for ordinary households to buy, whereas teenagers as a particular group were always curious about the new digital products. As a result, more and more teenagers got familiar with the Internet via the Internet café.

On April 18th to 21st, 1997, the National Informatization Working Conference was held in Shenzhen City, which passed the Ninth Five-year Plan for National Informatization and released the Perspective in 2000. The conference determined to establish Internet infrastructure in the whole country and proposed to set up the national Internet information center and Internet exchange center. According to the fifth report from CNNIC, in 1999, there were 747 thousand computers in China, which meant just 0.1 % of the whole population had physical access to the Internet. Among the Internet users, just 15 % claimed that they had an E-commerce and online shopping experience, and 87 % of the users were potential consumers (CNNIC, 2000).

In 1999, the Tsinghua University, which is the top science and technology university in China, set up a Computer Emergency Response Team (CERNET) to solve emergency cases. In the same year, the Internet was applied to university enrollment, and even though it was not popular among other industries, it was introduced to education first. More than two hundred colleges and universities recruited students online via CERNET's Colleges and Universities Enrollment System. In 2001, remote education or distance education was carried out by the CERNET. The Modern Remote Education called "Action Plan for Education Vitalization Facing the 21st Century" aimed to provide a life-long education system in China. Forty-seven colleges and universities joined the project in the beginning. Afterwards, more and more universities and private schools participated in the program as well. It is worth noting that the Internet network in China was designed with high standards at the beginning, which provided a solid foundation for the later network's extension, for instance, the remote education project built upon a high-speed transmission network with a capacity of up to 40Gbps (Internet, 2004; Xiang, 2009).

The government-oriented policy made significant progress. In 1994, the adopters were only hundreds of scientists, but five years later, in 2000, this figure reached 8.8 million. According to the latest figures by CNNIC (The China Internet Network Information Center set up June 3, 1997—their annual reports help the government formulate the next step of the Internet development project.), by the end of 2012, the number of Internet users in China was 564 million with an Internet penetration of 42 %. The big achievement was not only recognizable in the number of Internet users but also in the high-speed connections. As noted earlier, at the beginning the International Bandwidth was 300 bps, however, it surged to 2,643,660 Mbps by the end of 2016 (Telegeography, 2016).

From CNNIC[34] reports, we can find clues to the strategies employed by the Chinese government to diffuse the Internet. During ICT diffusion, the role of the government is undeniable. According to the report, Deng Xiaoping proposed that the diffusion of computers should start in childhood. After China joined the Internet, the State Council released two documents, namely *Establishing the First Experimental Plot of Computer Science* and *Consultation on the Requirement of Establishing Campus Network*. Both of these documents implied that the central government wanted to set up an Internet network in schools as soon as possible, and the central government was required to develop remote education to guarantee the people in rural areas have a chance to learn computer skills. Zhu Rongji, the former premier of the State Council, set up the first National Information Work Leading Group. On December 25, 1991, he proposed China should establish a market-oriented Internet economy and begin ambitious construction (CNNIC, 2017).

---

34 All the reports can be found from https://www.cnnic.net.cn/hlwfzyj/hlwxzbg/. Accessed February 6, 2017.

In order to achieve these targets, the government invested huge amounts, split the goal into different small tasks and planned each step. The first target was that every primary school and high school was obligated to design computer and Internet courses and every student should participate in the courses. In primary school, learning how to use a computer became a compulsory course. The Chinese government required all primary schools to put computer and information science into the list of compulsory courses, which meant the computer courses were as important as learning Chinese, math, science and society. By 2005, 90 % of all schools were planned to be equipped with computers and have Internet courses (CNNIC, 2017).

The second goal is the connect to each school (*Xiaoxiao Tong* 校校通) project, which aimed to equip 90 % of the schools with Internet facilities and computer classrooms. The central government wanted Internet access to be available for teachers and students within five to ten years. In China, government policy is regarded as the bellwether. As the government attached great importance to the development of the Internet, the Internet industry and market developed rapidly. It was estimated that in 2000 the market value of hardware and software reached 1.25 billion dollars even though the Internet was still at the threshold stage and in 2001 this figure reached 1.8 billion dollars. An investigation conducted in 2001 indicated that 52 % of the schools bought software for educational purposes, 38 % for course development and 32 % for teaching administration.[35]

On November 3rd, 2005, Premier Wen Jiabao[36] chaired the fifth meeting of the State Informatization Leading Group. As the contribution of the information industry to the Gross Domestic Product (GDP) amounted to 7.2 % in 2005, the meeting recorded the benefits that the Internet had brought to China, which contributed to 16.6 % of the economic growth. The export of electronic products accounted for more than 30 % of total exports. Furthermore, the meeting also discussed other Internet issues, such as trends, existing problems, and the next steps of development. The meeting approved *The National Informatization Development Strategy (2006–2020)*, which formulated new policy based on six aspects: education, e-government, technology innovation, informatization, digital divide, and Internet security (CNNIC, 2017).

---

35 See http://www.etc.edu.cn/articledigest11/Chinese-edu.htm. Accessed December 15, 2014.
36 Wen Jiabao was the premiere of State Council between 2003 and 2013.

On September 7th, 2007, the report on the Internet Development in Rural China[37] was released pointing out that there existed a digital divide not only between China and developed countries, but also within China, e.g., urban/rural and east/west, both, having large differences in Internet penetration. The report pointed out that the next urgent task for China was to increase the number of Internet users in rural areas. On October 5th, 2007, the 17th National Congress of the Communist Party of China was held in Beijing. At the conference, President Hu Jintao[38] said the future goal for the Internet in China was to establish an industrialized, information-based, urbanized, market-oriented and internationalized society. When the television was introduced, it was regarded as a symbol of modernization. Then, the Internet became the next symbol of this kind (CNNIC, 2007).

After 2007, Internet development started to focus on the cyber environment, especially Internet culture. For example, laws for supervision and administration over the E-commerce market, e-government, and intellectual property rights were released. On July 9th, 2012, the State Council proposed a project called "Broadband China", which aimed to improve transmission speed, expecting it to reach 50 Mbps and 12 Mbps in cities and villages respectively in 2020 (Internet, 2004).

The Inkomi search engine did a language investigation, based on more than one billion unique Web pages in the 1990s. They found that English documents accounted for 87 % of the whole Internet contents, and most web pages on the Internet were in English. This result was confirmed by the Babel team who did more research in 1997. They proved that 84 % of all web pages were in English, followed by German, Japanese, French, Spanish, Swedish, and Italian, with all other languages below one percent (Norris, 2001). However, after 19 years, are there any changes?

Based on reports of the Babel team, in 1997, the popular languages did not include Mandarin. However, in April 2013, W3Techs investigated one million of the most visited websites and concluded that almost 55 % of the most visited websites were in English, followed by Russian, German, Spanish, Chinese, French, Japanese, Arabic and Portuguese. The number of non-English pages was rapidly expanding. Among the non-English web pages, Mandarin ranked

---

37  The Chinese name is 2007年中国农村互联网调查报. The report can be found on the website of CNNIC. https://www.cnnic.net.cn/hlwfzyj/hlwxzbg/index_4.htm. Access December 15, 2014.
38  Hu Jingtao was the premier of State Council between 2003 and 2013.

number 5. According to the statistics of Internet World Stats (IWS), two years later, in 2015 the Chinese language jumped into second place.

In Fig. 3-7, the blue line indicates the Internet penetration of one language, and the orange line shows the users of one language as a percentage of the world population. Among the top ten Internet languages, users of other languages have better performance in participating in the Internet than Chinese and French users. The number of Chinese Internet users has grown very fast from 1997 to 2015. However, there is still an opportunity for more people to join the Internet. The biggest problem in China is the *pinyin* obstacle, education differences, and dialects making it very difficult to diffuse the Internet among the older generations.[39]

Concerning the diffusion of the fast Internet, the effort of the government is one of the important factors that need to be considered. The Chinese government highlighted the task to diffuse the Internet in a national plan and designed policies and regulations[40] ahead of the real tempo of the Internet development, for example, Internet privacy regulations (2000), domain register regulations (1997, 2000, 2002, 2004), self-discipline regulations (2004), security regulations (1994) and so on. Referring to the frontier countries' experience, the government utilized a policy to guide the economic direction and to stimulate the development of the Internet.

The consequences of the Internet are remarkable, affecting not only the economy of China, but also the communication system of the society. E-commerce, conveyed by the information society, revolutionized the traditional market. On September 6th, 1999, the Ministry of Foreign Trade and Economic Cooperation and the Ministry of Information Industry held the E-commerce Expo in Beijing to discuss e-business for a long-term plan. Thereupon, *All in One Net* was provided by the China Merchants Bank, followed by similar products at other banks. The new online services from banks made Business-to-Business (B2B) trade spread pervasively. Among the B2B companies, the most famous and successful one is Alibaba Group, which is a private company established by Ma Yun in 1999. Now it has grown into an online empire that covers Business-to-Business online marketplaces, retail and payment platforms, shopping search

---

[39] These problems invoke serious consequences for Chinese society, the old generation represents a traditional culture and connection between past social memory and present social memory, I am going to explore more in the later chapters.

[40] See http://www.china.com.cn/chinese/zhuanti/192893.htm. Accessed February 22, 2016.

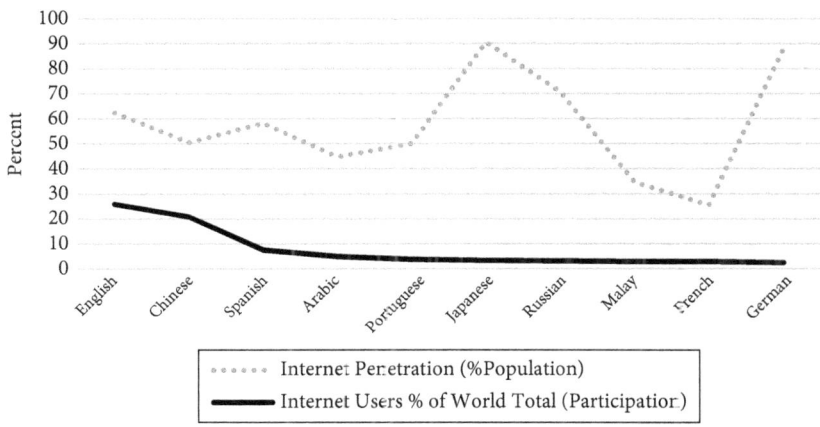

Fig. 3-7: Internet penetration and participation of the top ten languages used in the Web in 2015
Source: IWS (2015a).

engines and data-centric cloud computing services. The next step is its entry into the logistics industry. In 2012, two of Alibaba's portals together handled $170 billion in sales, and its revenue was more than the sum of eBay and Amazon.

On August 11th, 2005, Yahoo China announced their plan to transfer all business of Yahoo China to Alibaba. This was the first time that a global Internet giant company transferred business in China to a local Chinese company. Moreover, in 2007, the market value of Tencent, Baidu and Alibaba successively surpassed 10 billion dollars, which means Chinese Internet enterprises ascended to the list of the largest Internet enterprises in the world, and Tencent company became one of the top three Internet companies in the world Huge domestic Internet companies sprang up and many leading companies like Apple, Dell, and Nokia began assembling their products in China. Hence, plenty of grass-roots electronic companies work for these giant companies and some of them are starting to develop their own products.

In summary, it seems that China chose a unique road to boost the Internet, which does not mean that, as critics say, that the government strictly controls the Internet. The government has tried to adapt to the information age by increasing the number of online channels to communicate with the masses. For example, the microblog not only shapes the individuals but also changes the communication system between the government and the masses. On January

1st, 1997, the website people.com.cn became the first news propaganda website which represents the Chinese government. However, as the influence of microblogs increased, organs of the central government or local governments set up their own microblogs to better communicate with the masses. According to CNNIC, the year 2012 witnessed the burgeoning development of government microblogs. There were 60,064 government microblogs on Sina (*xinliang*新浪) microblog (*weibo*微博) and 70,084 government microblogs on Tencent (*tengxun weibo*腾讯微博) (CNNIC, 2017).

The role of the Internet is special. It not only boosted the computer industry, but also created incentives for the development of the mobile phone. On December 22nd, 2001, China Unicom Corporation declared in Beijing that the first-stage construction of the CDMA mobile communication network was completed as scheduled. On May 17th, 2001, China Mobile Corporation first launched the General Packet Radio Service (GPRS) service nationwide, and on November 18th, China Mobile Corporation and AT&T Wireless jointly declared that their GPRS international roaming service was officially put into operation (CNNIC, 2017). Based on these three events, China entered into the fast diffusion period of the Internet.

The prosperous Internet companies contribute to economic growth in China and the development of online shopping websites has changed the traditional way of selling goods and boosted other industries such as the delivery industry and transportation. However, recently, a bet from the top two entrepreneurs, namely Ma Yun, CEO of Alibaba Group and Wang Jianlin, head of Wanda Group[41] and also the richest person in China, have started a debate. Wang Jianlin said to Ma Yun that if in 2022, E-commerce occupies 50 % of the retail market, he will give 0.1 billion RMB (around 16 million dollars) to Ma Yun.[42] Fast E-commerce has been accused of being destructive to traditional "brick and mortar" entities, as more and more people choose to purchase online. People who cannot use the Internet cannot sell or buy products in real shops. The expanding online shops squeeze the space left for real shops, threatening their survival in the marketplace. The consequences of the fast development of E-commerce need to be re-evaluated and observed in the long term.

---

41 Dalian Wanda Group Co Ltd is China's largest commercial property company. It has many real estate projects. In recent years, it has started to focus on the film industry and is the world's largest cinema chain operator.
42 See http://finance.sina.com.cn/chanjing/gsnews/20131212/023917606582.shtml. Accessed March 22, 2016.

## Mobile Phones

After the era of newspaper, radio, television and the Internet, the smart mobile phone is regarded as the fifth generation medium that shapes the near future. Mobile phones, especially the 3G and 4G smartphones, not only have functions of the other four media, but also have better user interaction. Similar to the Internet, the mobile phone obscures differences between information administrators, transmitters, and receivers. Portable terminal users can publish news easily and the news can be spread fast. In the information society, everyone can be a news editor and receiver.

The mobile phone appeared in China in the late 1980s. When Guangzhou, a southeast coastal city in China, first introduced the Motorola 8500, China's media turned a new page. The mobile phone could only be used in Guangzhou city at that time, and the roaming service was only available after Beijing, Shanghai, and Shenzhen had mobile phone connections.[43]

The genesis of the mobile phone spread fairly slowly at the beginning. High prices and high fees scared many people away. The mobile phone represented the social class of the users. The mobile phone in Mandarin was called *dageda* at the beginning, which means big brother. The mobile phone was very heavy with a long antenna at that time and it looked extremely prestigious. The name of *dageda* suggests that people who could afford the mobile phone were bosses, leaders, and people who had a lot of money However, after experiencing Mao's period, people such as entrepreneurs and officers who might be potential customers to buy mobile phones were reluctant to show their wealth, authority and status. The symbol that mobile phones represented became an obstacle for diffusing mobile phones because of the specific social circumstance.

At that time, *xiaolingtong* emerged. In 1997, *xiaolingtong* was first used at Yuhang, Zhejiang province, which is a small city in east China. *Xiaolingtong* is the Personal Handy-phone System, which was prevalent in China before 2009. The cheap price even though without roaming service attracted many users. *Xiaolingtong* opened a door for spreading mobile phones and also occupied the mobile phone market. Ten years after the quick spread of *xiaolingtong*, it was substituted by standard mobile phones, which became the most popular communication device in China. The users who don't use roaming services frequently prefer to use *xiaolingtong* (Wang, 2008).

---

43 See http://wenku.baidu.com/view/6af55d00a6c30c2259019eef.html. Accessed July 16, 2015.

The reasons why *xiaolingtong* emerged and occupied most of the mobile phone market could be explained by the government-oriented telecommunications reform. After the third telecommunications reform, China Telecom Corporation and China Netcom cooperation lost market share in the mobile phone field. However, at the same time, the government did not decide which company was better to run licenses of the third generation (3G). Hence, the Ministry of Industry and Information Technology asked *xiaolingtong* operators to accept the responsibility to transfer *xiaolingtong* users into mobile phone users when the decision was made. In addition, the telecommunications field was a high-profit industry, and the mobile phone services providers, such as China Mobile Corporation, China Unicom Corporation, etc. were hardly able to reach a consensus in terms of the mobile phone fees. As a result, *Xiaolingtong* entered the market easily.

*Xiaolingtong* succeeded to persuade the majority of consumers to use it. Later on, it transferred the subscribers into mobile phone users, which bridged the divide between the fixed-phone and the mobile phone. After consensus among the telecommunications companies was finally achieved, the age of *xiaolingtong* ended on January 7th, 2009. The Ministry of Industry and Information Technology announced three companies would have a 3G license. To recapitulate, the China Mobile phone used TD-SCDMA; the China Unicom Corporation used the CDMA; and the China Telecom Corporation used WCDMA. Since then, *xiaolingtong* exited the stage of history. The Ministry of Industry and Information Technology prohibited *xiaolingtong* from developing their services in Beijing, Tianjin, Shanghai, and Guangzhou in order to support the mobile phone. At the same time, the Ministry of Industry and Information Technology claimed that the frequency band used by *xiaolingtong* was being removed from the grid of telecommunications networks by the end of 2011, which demonstrated the task of *xiaolingtong* was ended and China entered the era of the smartphone (Kong & Jin, 2003).

It is worth to note that, according to Ministry of Industry and Information Technology of China (MIIT), from the beginning of telecommunications services for mobile phones until they became ubiquitous was only 13 years; compared to the diffusion of newspaper, radio, and television, the mobile phone life cycle is much shorter. Fig. 3-8 shows the number of fixed and mobile phone users from 1998 to 2012. As shown in the chart, in 2003 mobile phone users exceeded fixed-phone users, becoming the most popular communication medium in China. At the same time, the CDMA mobile communication network and roaming service started to be used in 2002, which could explain the reason why the number of mobile phone users increased dramatically in 2003.

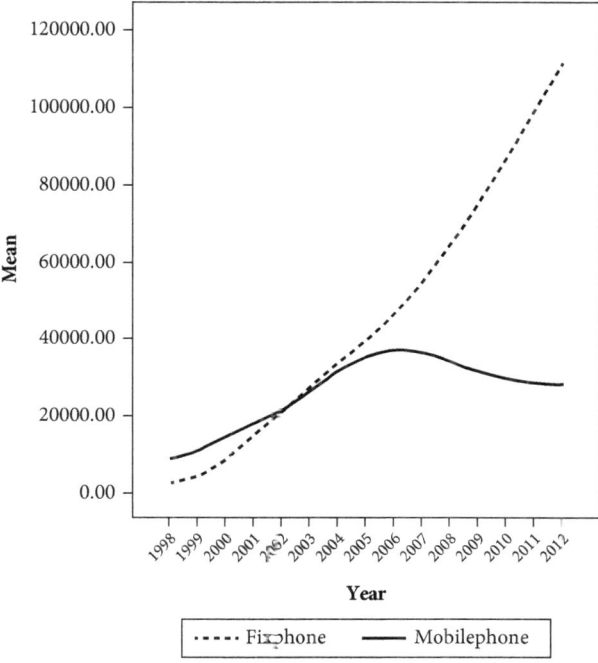

**Fig. 3-8:** Users of fixed and mobile phones from 1998 to 2012
Source: MIIT (2016).

There are several reasons attributable to the fast development of the cell phone. Firstly, in the global context, when mobile phones became prevalent in China, the world also experienced the same situation. The development of the mobile phone in China follows the trend of the rest of the world. Secondly, the development of one technology is always intertwined with other technologies. When mobile phones started to become widely used, other communication hardware and software were launched on the market, which supported the distribution of mobile phones within a short time and made it much easier. Thirdly, after several decades' development, non-mobile phones had established a systematic grid. Based on the facilities of fixed-phones, the mobile phone spread among entire countries within a short time.

After elucidation of the development of the media, differences in the distribution of old media and new media become apparent. Contemporary China is influenced by ICT, the new media is gradually replacing the old, and the transition is accompanied by a transition of the entire society. Developed countries are regarded as frontier countries in terms of innovation diffusion, and they give important hints to the laggard countries.

In a social disorder period, the government prefers to use listening media devices rather than other types of media. The traditional mass media, such as the press, radio and television are easier to control than the new mass media, for instance, computers and smartphones. The development of radio suggests how a government may set up preferential policies and selectively develop one medium during a certain period.

## Conclusion

History furnishes numerous cases of success and failure of innovations and provides useful insights to understand human society. It has explained why some sciences or technologies are more easily adopted in one country or a specific period while others are not. It has also explained the differences in diffusion patterns of media and the roles of media in traditional society and modern society (I. E. Fang, 1997). By using a historical approach to introduce the old media and new media in China, it suggests how the diffusion process of media happens in China.

The diffusion of ICT in China is politically oriented and the opinions of leadership play important roles in the diffusion process. ICT began entering China one hundred years ago, but it started to diffuse faster after the 1980s. The Chinese government referred to other frontier countries' experience to design the policy, which guided and stimulated ICT diffusion. The diffusion of ICT has gone hand in hand with policy preferences, demonstrating that a laggard country can base their policy on the consequences of Internet diffusion from frontier countries. No one could predict that the media could make such progress under tight control. The diffusion of ICT balances the information inequality between government and the masses that users can be information producer and receiver. During the modernization process, technologies are not equally developed; the invention of technologies always intertwines with society to make social changes. Hence, both the government and individuals adjust themselves to the social change.

The social condition is an important factor that decides adoption of the media, and media tends to adjust themselves according to the social condition. The traditional mass media, such as the press, radio, and television are easier to control than the new mass media. In an unstable social condition, the government's prior use of the loudspeaker and radio proved that the listening mass media are more popular in a social disorder period than other media. For instance, the radio contributed to maintaining a stable society and made sure that every person got the same information, policy or propaganda from the central government. In addition, the social condition also largely decides the distance between leaders and the masses. In a social disorder period, government leaders will be more secretive compared to a period of social stability. The social condition shapes how media develop, meanwhile, the development of media influences the change in society. The comparison of the Mao period and post-Mao period suggests that during times of social disorder, the government tries to control the media and takes measures in order to keep society stable.

# 4 Is the Diffusion of ICT in China Special?

## The Diffusion Curve of ICT in China

For the old media, users are getting information passively and the way these media work is simple. Compared to the new media, they are easier to control. Since the spread of the new media, the old media is gradually fading away as a communication system. Demonstrating China as a new modernized society, the penetration of newspaper, radio and fixed-phone has decreased faster than the frontier countries, such as Denmark, Norway, etc.

The developed countries, such as United States, Australia, Germany, and the Nordic countries (the Netherlands, Denmark, Norway, etc.) are at the front of the technological revolution. The non-Internet users in these countries are "information want-nots"—the reason they do not use ICT is not that they cannot afford it, but that they do not want it (DiMaggio & Hargittai, 2001; Kamarunzaman et al., 2011; Lu Wei & Zhang, 2006). However, the situation in China is different, many still belong to the "information have-nots". Considering the information skills, most of the information users in China only have the basic level to use the Internet and a huge divide exists among users. This crucial aspect contributes to the digital divide in China as an urban-rural gap, not just on the physical access level but also on the usage level. As the cyber-pessimists said, the early adopters who already benefited from ICT will maintain their relative lead in the digital economy (Norris, 2001).

The early adopters in China were mainly living in the urban areas or on the east coast of China, while the rural regions and areas in the west were on the edge of the information society. As the living standard improves, the author assumes that the gap between rural and wealthy regions will narrow. In terms of the usage of ICT in China, to what extent can China take advantage of the introduction and diffusion of innovations and what is different in the diffusion process between China and other developed countries like the United States, the Netherlands, Germany, the United Kingdom and so forth?

Based on this assumption, the eastern, middle and western parts of China have different living standards and industrial levels, hence, the author assumes different provinces with different GDP per capita will have different diffusion patterns in regard to media penetration, especially the information and communication technologies.

As China can be divided into three regions, namely the eastern, middle and western parts, I will categorize the provinces according to the Chinese central

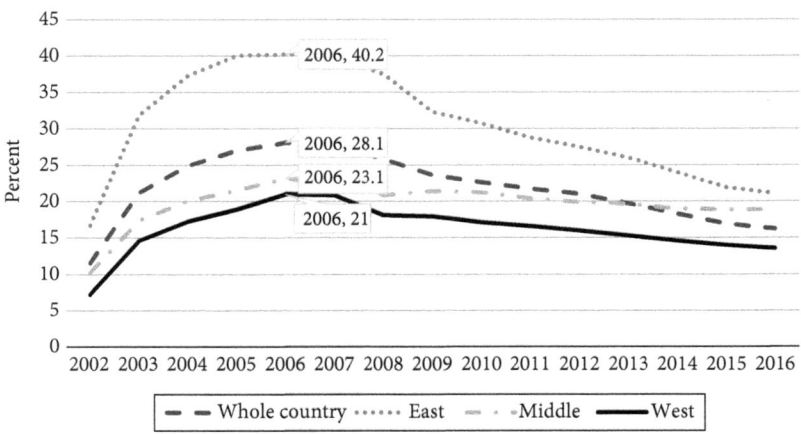

**Fig. 4-1:** The trend of fixed-phones in China from 2002 to 2016.
Source: MIIT (2016).

government's classification to divide the provinces and cities. The eastern part includes these 11 provinces and cities: Beijing, Tianjin, Hebei, Liaoning, Shanghai, Jiangsu, Zhejiang, Fujian, Shandong, Guangdong and Hainan. The middle part is composed of these 10 provinces: Shan'Xi, Neimenggu, Jilin, Heilongjiang, Anhui, Jiangxi, Henan, Hubei, Hunan and Guangxi. In addition, these 9 provinces make up the western region: Sichuan, Guizhou, Yunnan, Xizang, Shanxi, Gansu, Qinghai, Ningxia and Xinjiang.

In terms of the diffusion of the fixed-phone, Fig. 4-1 shows that its spread throughout the whole country peaked in 2006 and then started to decline. No matter whether the eastern, middle or western part of China, all reached the saturation point in 2006. Concerning the mobile phone, Fig. 4-2 indicates that the diffusion of mobile phones increased very fast from east to west. The eastern part leads the trend as the middle and western parts follow. Since 2002 in China, the diffusion of mobile phones increased until a peak in 2013, and then entered a plateau period suggesting saturation. It also suggests that from 2002 to 2007, the diffusion curve of the mobile phone in central China and western China overlapped, but since 2008 central China developed faster than the western part, which means the incubator with a higher industrial level tends to develop and adopt new innovations more quickly.

To further test the relationship between industrial level and the diffusion of ICT, I put 31 provinces and cities together to see the rate of ICT increase per

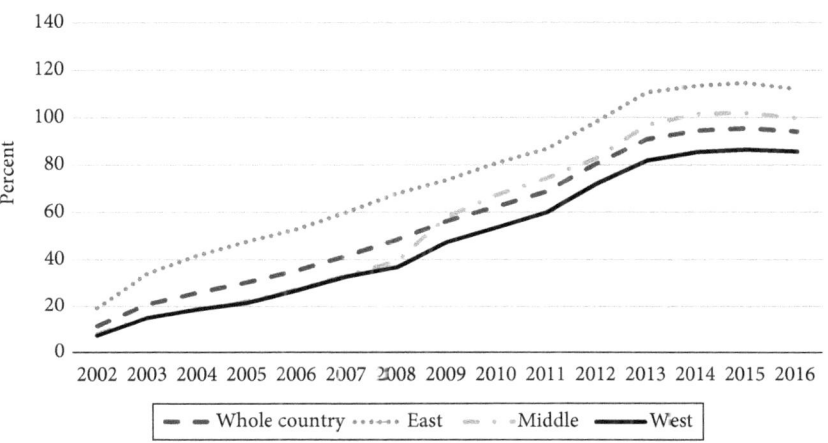

**Fig. 4-2:** The trend of mobile phones in China from 2002 to 2016.
Source: MIIT (2016).

year. The data, from the Ministry of Industry and Information Technology of the People's Republic of China from 2002 to 2016, analyzes the diffusion rate of fixed-phones (Fig. 4-3) and mobile phones (Fig. 4-4) and proves that cities and provinces with high industrial and living standard levels can diffuse the technology within a short time.

Compared to the middle and western parts, the eastern part of China has the potential to improve the diffusion of media to a high level within a short time. The old media were gradually replaced by the new media, though they were not fully developed in the laggard places. Based on the diffusion of ICT in China, the rate of the fixed-phone in eastern China is higher than in the middle and western areas. As mobile phones became more available, the fixed-phone diffusion rate of all three regions slowed simultaneously—the laggard areas chose the new ICT rather than attempting to catch up to the eastern level of the old ICT.

Concerning Internet diffusion, analysis of the GDP per capita of the provinces and cities suggests that Internet penetration is correlated with higher GDP. According to social-economy development reports, provinces and cities such as Shanghai, Beijing, Zhejiang, Guangdong, Jiangsu, etc. have the leading Gross National Income per capita and GDP per capital in the last five years along with relatively higher consumption levels compared to the other provinces (see Fig. 4-5).

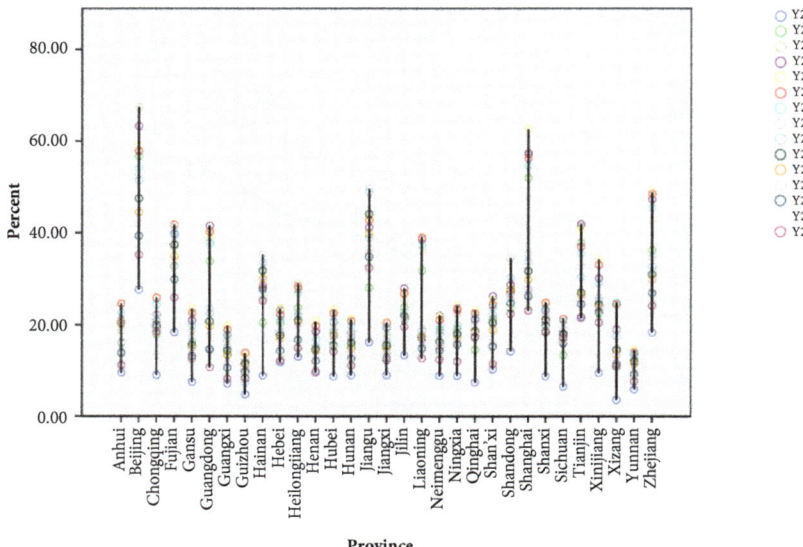

**Fig. 4-3:** The fixed-phone diffusion rate of each province from 2002 to 2016.
Source: MIIT (2016).

After analyzing the diffusion curves of fixed-phone, mobile phone, and Internet, it suggests that provinces and cities with high GNI and GDP per capita are the frontier places in China that can diffuse new innovations quickly and popularize a technology to a high level. When the mobile phone starts to diffuse, no matter fixed-phone reach to the saturation level or not, the penetration spread slows down.

## A Comparative Analysis of ICT

To understand ICT diffusion and the factors which influence the diffusion dynamics, the following discussion examines China's position with respect to 24 other countries. The countries are chosen according to the GDP and GNP as well as their comprehensive performance: South Korea, France, Germany, UK, Canada, Japan, United States, Denmark, the Netherlands, Spain, Estonia, Russia, Ukraine, United Arab Emirates, Brazil, China, Chile, Mexico, Malaysia, Indonesia, Thailand, South Africa, Mongolia, Philippines, and India. The data are gathered from the International Telecommunications Union (ITU), which is the United Nations specialized agency for information and communication technologies—ICT. These countries include developed countries and developing

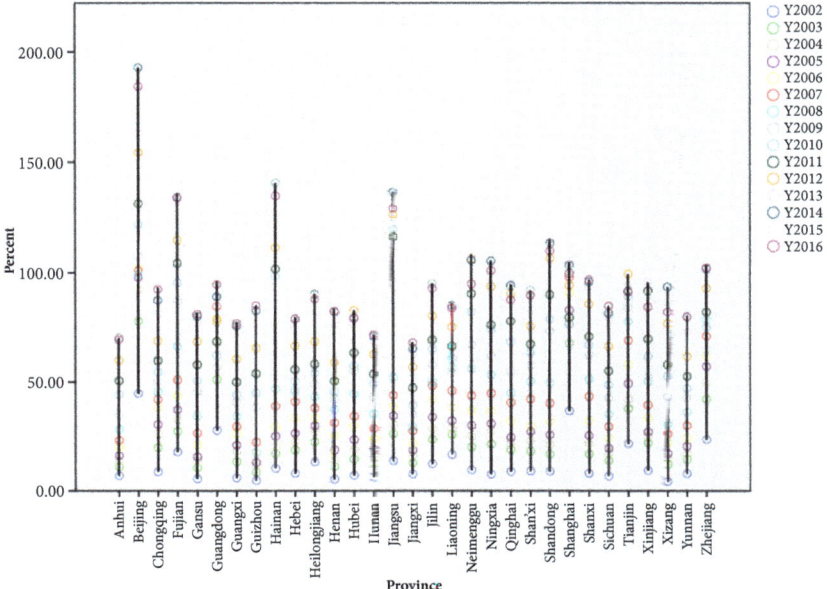

**Fig. 4-4:** The mobile phone diffusion rate of each province from 2002 to 2016. Source: MIIT (2016).

countries, and most of them are very active in the international arena. Looking at the diffusion of ICT in the context of China will be very helpful to build a general impression of the diffusion curve of ICT.

Cesare Marchetti, after studying energy substitution and media concluded that the diffusion of technologies could be reduced to time, spatial considerations, and speed of the introduction of each new competitor. All the rest was a consequence, even 100 years later (Grübler, 1991). In terms of technology diffusion, Marchetti made the point that the rules of the game do not change very much (Ausubel, 1991). Based on this assumption, in the next paragraphs, I will choose the fixed-phone, mobile phone and broadband to analyze their diffusion patterns and respond to the argument from Marchetti.

## Fixed-phones

Fig. 4-6, combining ITU data from the 25 countries listed above, suggests that there are mainly three patterns that could describe the diffusion curve of the fixed-phone:

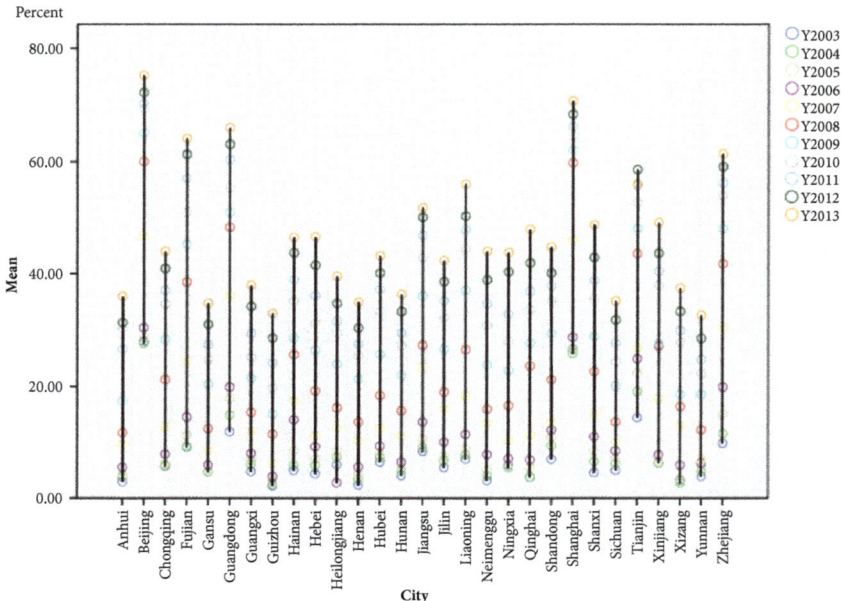

**Fig. 4-5:** The diffusion of the internet in China from 2003 to 2013.
Source: NBSC (2016). Note: Internet diffusion includes computer terminal and mobile phone terminal.

**Declining pattern** — with high adoption and dramatic decline that is demonstrated by Nordic European countries such as Denmark, Netherlands, Sweden, Norway, etc. and North American countries such as Canada and the United States. Most of these countries reached their highest penetration at more than 60 % in 2000; afterwards, it gradually decreased to around 40 %. Denmark even declined to 33.22 % by 2014, which is a very interesting phenomenon, since it is the most developed country that adopted the fixed-phone very early.

**Relatively stable line** — this trend is represented by Germany, France, and UK, whose penetration fluctuates between 52 % and 67 %. Germany and France peaked at 65.35 % and 65.09 % in 2005 and 2009, and then the penetration rates started to decline to about 58 %. However, compared to other developed countries the rate of fixed-phone subscriptions per 100 inhabitants is still very high. Other examples are Japan and Korea, both of which keep a stable diffusion rate. In 2000, the fixed-phone subscription in Japan was 49.28 %, however, 14 years later, in 2014, the rate is 50.09 %; it seems the diffusion rate does not change.

Similar to Japan, South Korea's line was almost flat—the diffusion rate only increased 2.19 % from 2000 to 2014.

**Normalization pattern** — the whole trend keeps a relatively low adoption rate, which is represented by most developing countries. The diffusion in these countries could be imagined as a bell curve with high $\sigma^2$. These countries experience a slight period of increased growth before they start to drop slowly. For instance, China and Brazil, after a short growth period, plummeted to 20 %, while countries like the Philippines kept a diffusion rate of around 4 %, and India is even lower at 3 %.

Analysis of the diffusion figures of the sample countries from 2000 to 2014 suggests that culture, history, industrial level, territory size, etc. shapes their diffusion curve trend. That is, countries with similar historical-cultural backgrounds, ideology, territory size, industrial level and so on have similar diffusion curves patterns.

However, in viewing the holistic trend of the fixed-phone over time and space, an interesting phenomenon is that the diffusion of fixed-phones in developing countries does not increase over time until it reaches a tipping point and then starts to decrease, as would be expected. According to the data between 2000 and 2014, most of the countries crossed the tipping point between 2005 and 2008, except for a few that peaked before 2000. In recent years, fixed-phone penetration in developed countries is gradually shrinking, yet diffusion in developing countries also shows the same trend even though the saturation level was far from being reached.

The reason the penetration of fixed-phones decreased could be attributed to competition from mobile phones and the Internet. However, it is known that fixed-phone diffusion is closely related to a country's industrial level, as government organs, institutions, factories and so on, all need their fixed-phone number for better communication. The diffusion of fixed-phones is one of the important indices that reflects a country's economy. For the laggard countries, there is still space to improve the rate of the fixed-phone's adoption. However, the reality is that the developing countries are mainly following the trend of developed countries, though the penetration rate is at least 30 percentage points below the saturation level.

Considering the case of the fixed-phone, the curve of diffusion is different from cars, refrigerators, televisions, etc.—it seems the rules of the game have changed. On this basis, the diffusion of technology has its own curve, which is not decided by development status, since both developed and developing countries reach their peak fixed-phone penetration by 2005 and 2006 and then the diffusion curve starts downward. In Germany, for example, the fixed-phone subscriptions per 100 inhabitants was 61 % in the year 2000, afterward it reached to the highest point at 65.35 % in 2005, and then it slowly decreased to 56.89 %

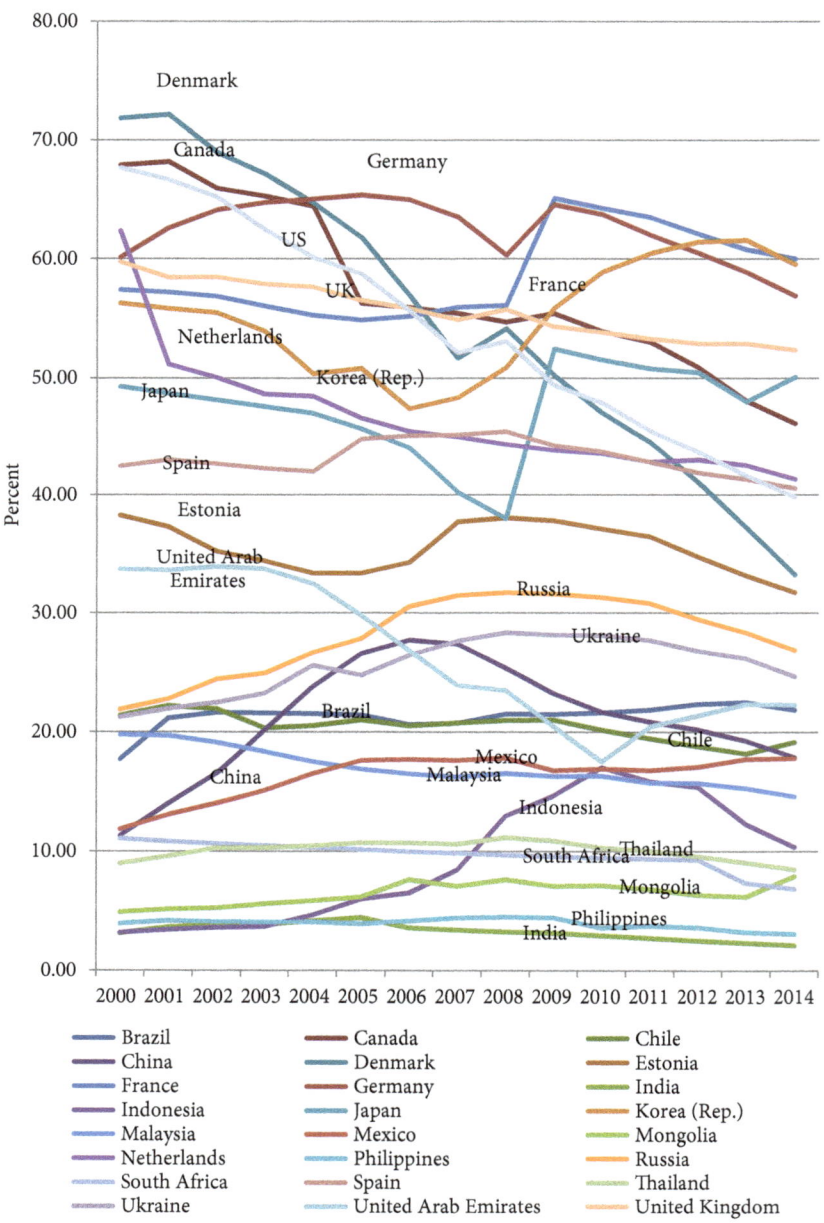

**Fig. 4-6:** The trend of fixed-phone diffusion between 2000–2014.
Source: ITU (2015).

in 2014. Similarly, in 2000 the diffusion rate of fixed telephones in India was 3.11 %, which peaked at 4.45 % in 2005, finally dropping to 2.13 % by 2014.

Fig. 4-7 suggests that 40 % can be seen as the tipping point of the fixed-phone between developed countries and developing countries. The rates in developed countries are all above 40 %, the less industrial countries are between 30 % and 20 %, and the laggard countries are below 10 %.

In considering the diffusion of fixed-phones, it cannot simply be taken for granted that fixed-phone diffusion is determined by incubations or regional seedbeds. The reason can tentatively be attributed to the fact that media have their own trend, which harnesses each country or region, and the diffusion of media in each country will follow the mass media's self-adjustment. Each country might have a different diffusion curve and different domestic situation contributing to the variation in diffusion performance, but the mass media's self-adjustment has the most pronounced effect on the diffusion in each country. Similar phenomena are also proven by diffusion of the mobile phone and broadband. Questions such as how much one country can abet or retard the diffusion of medium and which factors contribute to the diffusion curve need to be considered. Next, I am going to explore the diffusion of the mobile phone and broadband in response to these offsetting tendencies.

## Mobile Phone

Unlike the fixed-phone, the diffusion of the mobile phone is increasing and filled with energy. Fig. 4-8 suggests mobile phone subscriptions in both developed and developing countries increased dramatically before 2012; afterwards, they started to grow slowly. The spread of mobile phones in most countries is over 80 %. By the end of 2014, there was not much of a difference between developed and developing countries of mobile phone subscribers per 100 inhabitants, and some of the developing countries reached 150 %, exceeding developed countries.

The countries with low fixed-phone penetration have had a good implementation in diffusing the mobile phone. In the case of fixed-phones, developed and developing countries have sharp differences when observing diffusion curves, however, the differences in mobile phone diffusion are not that obvious. A good example to explain this is India and Canada. In 2014, the number of mobile phone subscriptions per 100 inhabitants in India was 74.48, which is the lowest diffusion rate among the samples. However, in Canada the figure was 81.04. Along with India, countries such as Philippines, Mongolia, South Africa, etc., also have shown very promising pictures in terms of mobile phone diffusion.

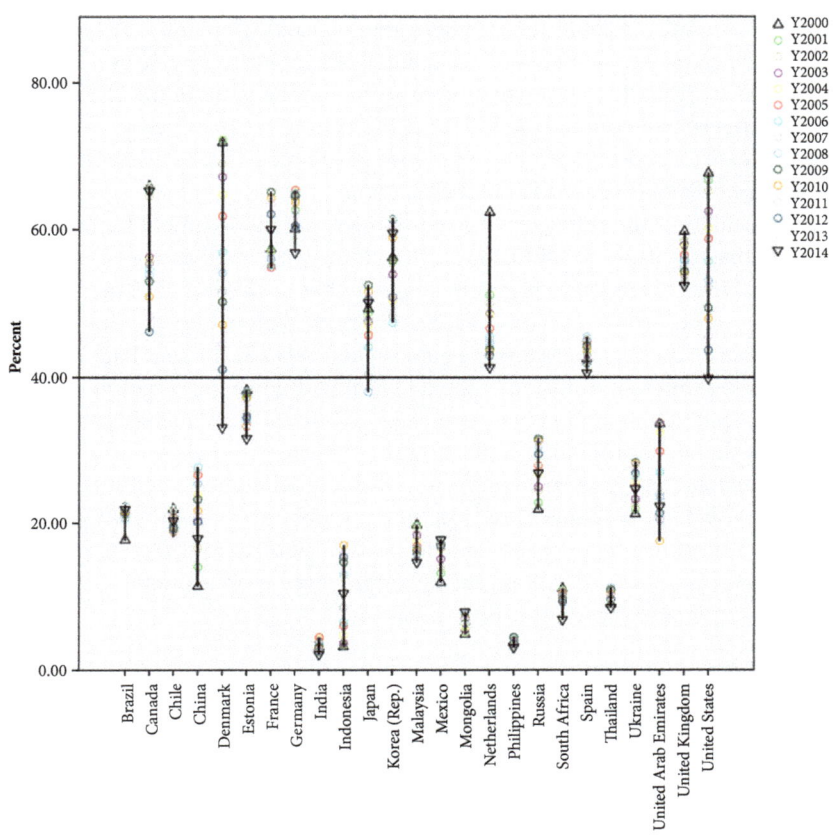

**Fig. 4-7:** The fixed-phone diffusion rate from 2000 to 2014.
Source: ITU (2015).

Only observing the data from recent years cannot distinguish the diffusion curves between early adoption countries and peripheral countries—a long-term trace is necessary to see the difference. By the year 2000, mobile phone diffusion in most developed countries had reached 50 % already and had increased steadily by less than 10 % every year afterwards. At the same time, the diffusion in developing countries was just at the beginning stage. According to the surveys of ITU between 2000 and 2014, the mobile phone diffusion curves in developed countries are similar. The most important distinctions of mobile phone diffusion exist in developing countries.

A Comparative Analysis of ICT 107

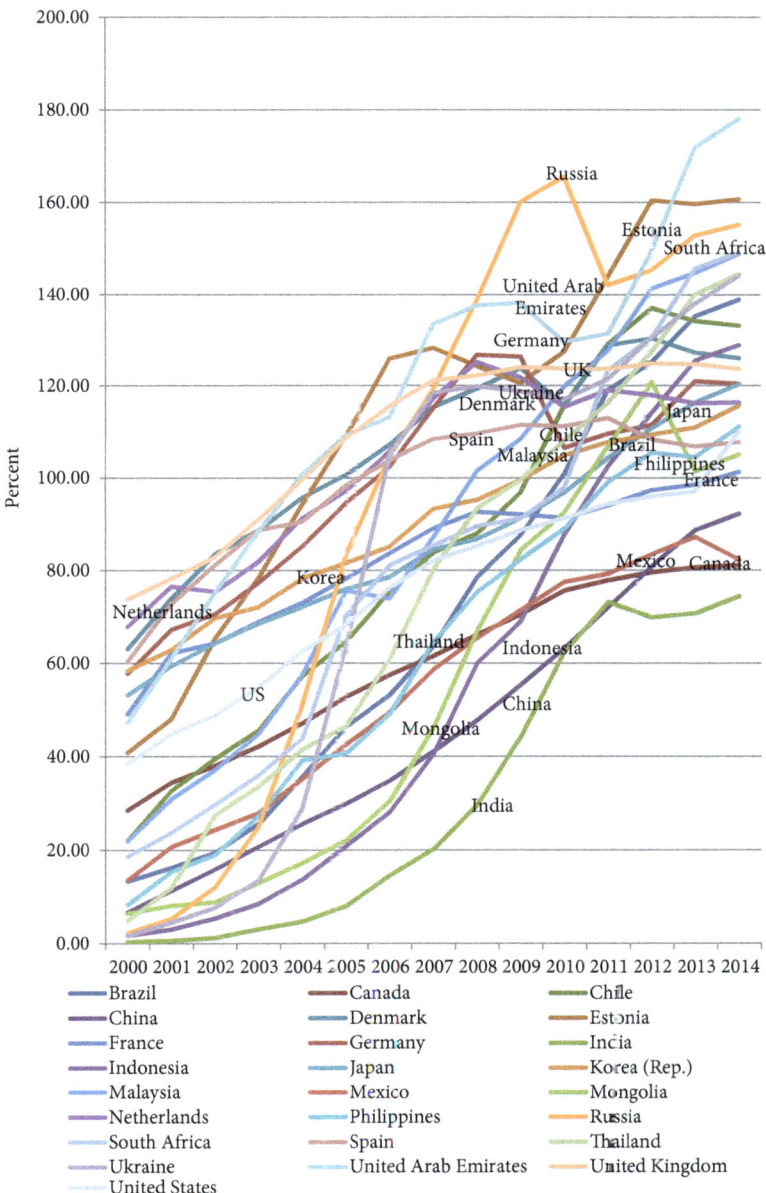

**Fig. 4-8:** The trend of smartphone diffusion between 2000–2014.
Source: ITU (2015).

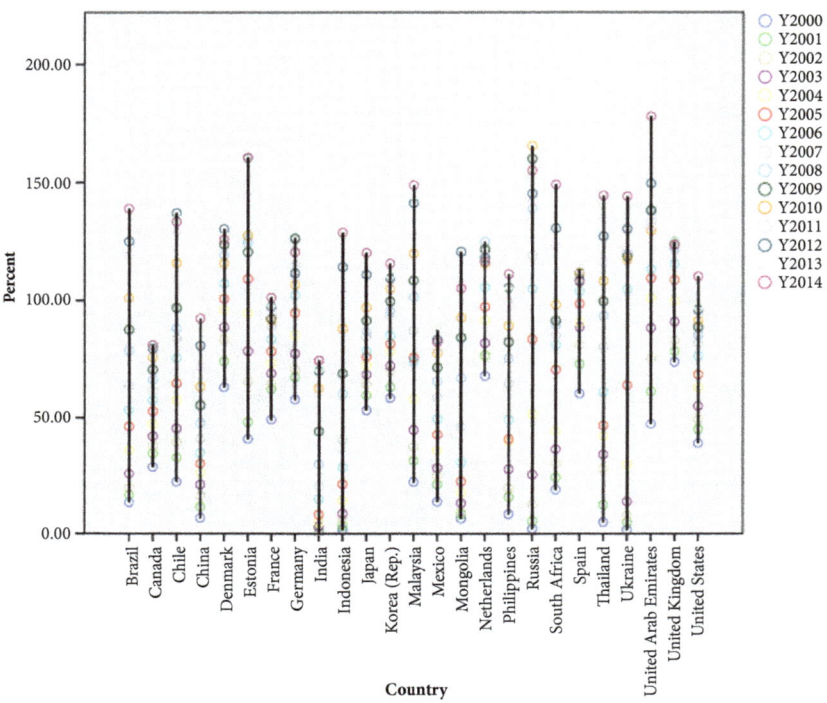

**Fig. 4-9:** The smartphone diffusion rate from 2000 to 2014.
Source: ITU (2015).

The diffusion type can be categorized into two types: smooth growth and leap growth. The first type is represented by China, Brazil and Mexico. In the past decades, the diffusion of mobile phones in China increased 5 % per year, as did Brazil and Mexico. The second diffusion type is illustrated by Russian, Ukraine, Estonia and the United Arab Emirates. In Russia from 2003 to 2006 the penetration rate of mobile phones increased by 30 % every year. Ukraine witnessed the most incredible growth since the diffusion rate was 29 % in 2004; however, by 2008 the rate surged to 120 % (see Fig. 4-9).

Understandably, as the price of mobile phones becomes ever lower and the multi-function ever greater, the smartphone increasingly shapes our lives. Communication technology has become part of our daily life, we use it not only for communication, but also for entertainment, self-teaching, GPS, photos, music, reading, shopping online, payment and so on. Huge progress has been made in both hardware and software, which are more accessible and usable for the majority of people. With easier access and multi-functions, the mobile phone has spread like wildfire worldwide. It is the first time the developing countries diffused one medium within a few decades to match or exceed the progress of developed countries.

## Broadband

A country's industrial level and comprehensive situation determine the diffusion of technologies. Mobile phone data from ITU indicates that in Western countries, for instance, the Netherlands, Denmark or Germany, Internet diffusion reached a relatively high level within a short time. Similarly, in Russia, the mobile phone adoption rate increased to 60 % just within two years.

However, the developing countries developed broadband relatively slowly compared to developed countries. The report from the ITU of broadband subscriptions per 100 inhabitants indicates that in 2000, except for Korea, most of the countries were at the same level. Since 2002, the developed countries suddenly accelerated the pace and improved the diffusion of broadband dramatically. For example, in 2002, the broadband diffusion in the Netherlands was 7.3 %, and by 2006, the rate surged to 31.7 %. However, the diffusion in India only increased 0.1 between 2002 and 2006. This phenomenon shows that the industrial countries have more potential to accelerate adoption of one technology at the beginning and middle stage. Since the development of technologies such as broadband depends on a country's industrial level and user's education background, it is difficult for an emerging nation to improve them within a short time (see Fig. 4-10 and Fig. 4-11).

To further analyze, there is an obvious line to distinguish the trend between the two groups of countries (see Fig. 4.10). By calculating the average rate of each year, namely from 2000 to 2014, if 1999 is regarded as the origin, 2000 will be equal to 1 (2000=1, 2001=2......2014=14), and according to the regression of curve estimation, we get a quadratic curve equation as below:

$$y = -0.0384 x^2 + 2.168x - 2.36$$

J. A. Van Dijk (2013) postulated that four factors influence the diffusion of ICT, namely motivation, physical and material access, digital skills (content creation, strategic, information/communication, operational, etc.) and usage (J. A. Van Dijk, 2013). The diffusion of broadband is more complicated compared to the fixed telephone, mobile phone and television sets. It largely depends on a country's industrial level, policy, individuals' background and motivation, etc.

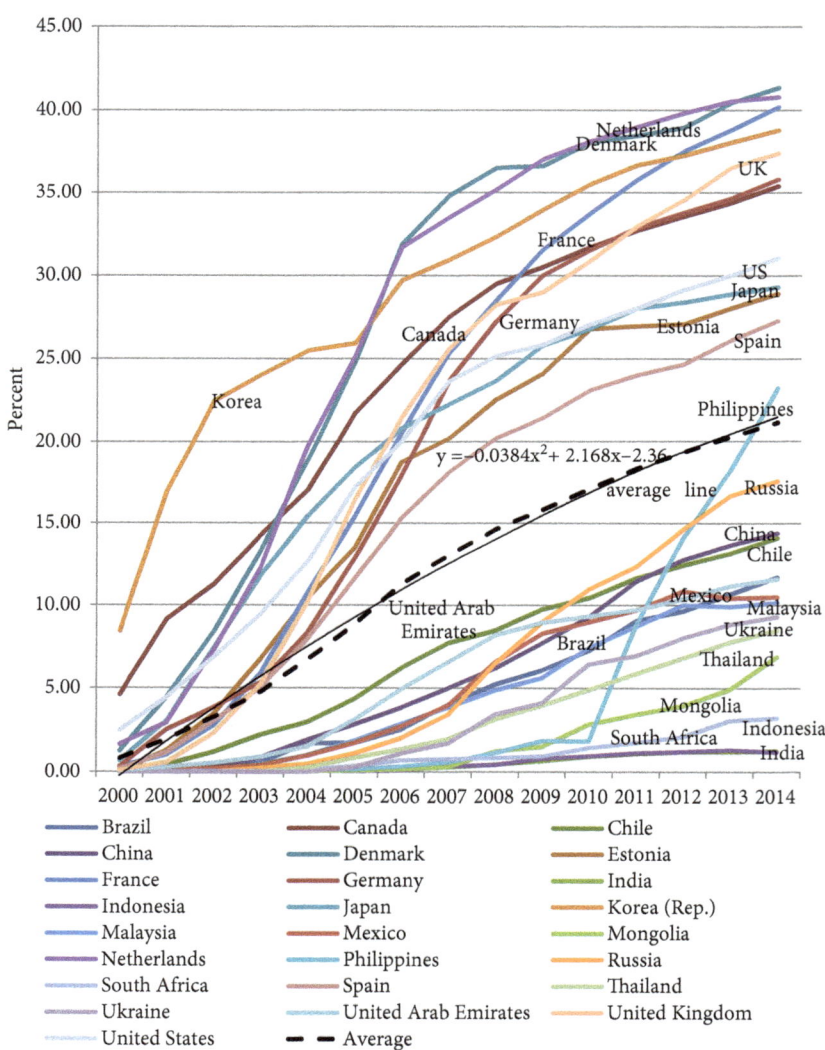

**Fig. 4-10:** The trend of broadband diffusion between 2000–2014.
Source: ITU (2015).

A Comparative Analysis of ICT 111

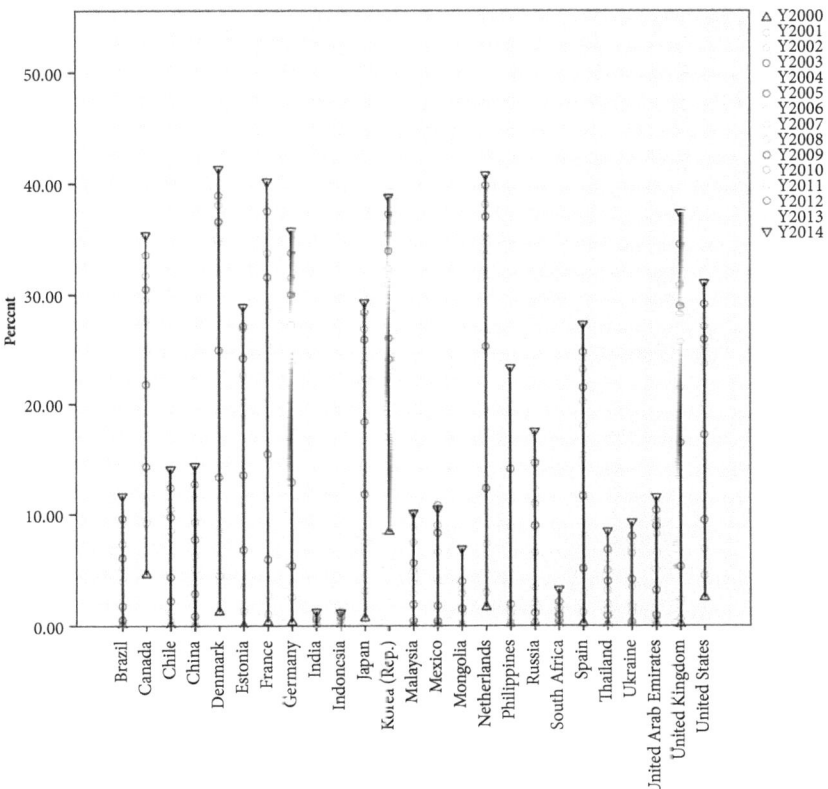

Fig. 4-11: The broadband diffusion rate from 2000 to 2014.
Source: ITU (2015).

When the diffusion of innovation is analyzed, the familiar S-curve of the adoption of innovations is widely applied. In the twentieth century, Sorokin (1941) drew upon classic theories of technological diffusion, and his concepts were later developed by communications scholars Elihu Katz and Everett Rogers (Katz & Lazarsfeld, 1955; Rogers, 2010). By analyzing the diffusion of the media, e.g. telephone, television, and radio, they found that the trend of the diffusion of these media always follows an S- (Sigmoid) shaped pattern. Because of high cost and a skills barrier, new technologies often have to experience a slow rate of initial adoption at the beginning followed by a normalization pattern. As markets

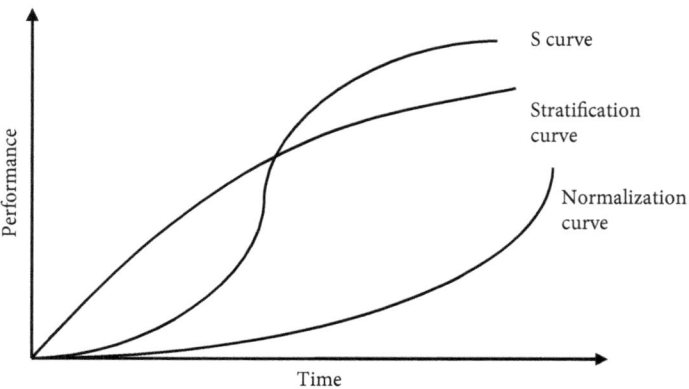

**Fig. 4-12:** The diffusion curves of innovations.
Source: Adapted from Rogers (2010).

expand, new technological devices will become cheaper and easier to use, until finally, penetration will reach the saturation level and laggards will catch up and the digital gap will be decreased (Rogers, 2010).

In order to understand the diffusion of ICT better, scholars try to use models to explain how innovations diffuse. Alexander Van Deursen and Van Dijk (2011), after studying the Internet diffusion of the Netherlands, concluded that the normalization model can be applied to explain the Netherlands and other rich countries; the stratification model is more fit for developing countries and gives a better reflection of the current and probable coming situation (A. J. Van Deursen, van Dijk, & Peters, 2011).

However, according to Fig. 4-10, it seems that the diffusion curve of broadband is different from the expectations of scholars. In the case of broadband from 2000 to 2014, the stratification model fits developed countries whereby the normalization model shows the trend of developing countries, the opposite of the predicted curve.

In terms of using one pattern to conclude the diffusion of ICT, the sample countries suggest that there is no one pattern that can apply to the diffusion of one technology from the beginning to the end. Each ICT has differential diffuse patterns. The traditional media, from 2000 to 2014, experienced a plateau period and then began to decrease. For the new media, the main trend is still increasing.

Analysis of the diffusion curves of the fixed-phone, mobile phone and broadband suggest that similar history, culture and industrial background tend to show similar technology diffusion curves. Comparable countries such as Japan and South Korea, Germany and France, Norway and the Netherlands, United States and Canada, etc. have a similar routine of technology penetration, albeit the curve fluctuation is not exactly similar but the whole trend shows a convergent penetration procedure. The diffusion curves of fixed-phones, cell phones and the Internet indicate that the developed countries or the high-per capita GDP countries can use a relatively short time to improve the diffusion of a technology which needs more investment in innovation compared to the low-per capita GDP countries. For technologies with cheap physical access and which need less investment in infrastructure and research & development, there is not a great difference between the diffusion patterns. When the same analytical method is applied to China, since east, middle, west China have different industrial levels, it also indicates that the laggard places follow frontier places and the industrial level of incubators decides whether they have the potential for fast diffusion.

## Conclusion

The introduction of the development of media in China and use of a comparative approach to understand the diffusion of ICT in China suggests that the diffusion of ICT in developing and developed countries could have different patterns regarding different technologies in a certain period; i.e., the diffusion of innovations cannot be simply generalized in one or several patterns since different countries have various levels of performance closely related to the historical background, culture industrial level, policy, social conditions, and so on.

In the beginning, I regarded the society of China as a fragmented society, as the eastern part, middle part and western part of China have big differences in terms of GDP, industrial structure and living standards of individuals. After analyzing the diffusion pattern of ICT in eastern, central and western China, as well as these selected countries, it suggests the diffusion curve of ICT in China shares a similar tendency with the selected countries. The different development of ICT closely relates to the unequal economic performance between developed and developing areas. Except that the industrial level, culture, history, territory size, etc. are important factors which shape the diffusion trend. Countries that share these factors tend to show similar diffusion curves patterns.

The performance of China's ICT diffusion is optimistic. However, there is an information gap between different regions, such as in the eastern and western parts of China, and between the urban and rural areas. Therefore, the unbalanced development may lead to an increase in the rich and poor gap.

# 5 Losing and Rebuilding the Social Trust System

Nowadays, the central Chinese government has less control of the available information. The new media are considerably different and every participator can be an information publisher and receiver, which means information belongs to the whole society, and individuals have personal physical access to information. In China most provinces and cities have a mass media monitoring center, the contents of traditional media were mostly under the control of the government and the government decided which kind of content could be published and which kind of content is forbidden. However, even though most of the content in the media can be supervised, the speed of information transference is quicker than the monitoring speed, sometimes allowing the so-called "forbidden news" time to exist in the public sphere before deletion.

The social condition shapes the usage of media and the usage of media also influences the social condition. In a mature and constitutional society, the media work as a social security valve, which helps the society release tensions of social stress. The main problem is sometimes journalists who report this news and want to arouse the masses attention may find the events too dull, so they exaggerate the reality or even report the fake news.

Since 2005, violent attacks on kindergartens have increased rapidly. Most of the criminals hate society and government and are transforming their anger into brutal knife attacks on innocent children. The media hyped the violence and even described in detail how the criminal committed the attack. This kind of news in such detail was harmful to society and spawned copycat attacks, several occurring within months throughout China.

As the widespread of ICT, conflicts between individuals and their local government are also increasing. When conflicts happen, the media likes to publish this kind of news to attract the attention of readers using the sensational and fake information to generate excitement. The media tries to describe the individual as weak and the local government as powerful, then the masses feel mercy for the individual and stand on the same side as the "victim"; however, the local governments, worried that this kind of impression will jeopardize the leader's political career want to resolve the issue as soon as possible to reduce the attention of the public. As a result, no matter right or wrong, the government compromises to individuals and agrees with the requirement that they asked.

The reportage on conflict makes individuals believe that the demonstration in front of the door of the government will help solve problems, so they choose this kind of way to achieve their targets. Sometimes, before they start to demonstrate in front of the door of their local government, they invite journalists to come along and pay them to report the process to arouse the attention of the masses. Consequently, to invite the masses to be a judge by publishing the conflict on mass media has become a routine to solve problems.

The sensational news makes the masses believe that violence and conflict can raise social attention. In order to raise attention from the public, criminals commit many horrible attacks and brutal crime on innocent people, such as igniting buses, attacking kindergartens, hospital massacres, and so on. The media releases of this kind of information do not help the society, on the contrary, it brings risks to the society. The authority of the mass media gets to the lowest point and at the same time, the influence of the mass media reaches its peak. During Mao's period, the media represented the authority and the central government. However, the current media has returned to the real role—information providers. The positive function, as far as can be seen, is to provide an access for the individual's voice and play a role of police and judges of the society. In order to better to understand how the government and the masses adjust themselves in the information age, I will introduce two of the important social issues that changed the communication system in the post-Mao age and caused wide debate online, they are the SARS virus and Guo Meimei.

## The SARS Virus

The reaction of the government, media and the masses when SARS[44] happened suggested the old media policy could not fit the information society, invoking serious results and triggering whole society anxiety. The SARS outbreak reflected that when a public crisis happens, the media has a crucial influence on the society.

On the 16th of November, 2002, the virus called SARS first appeared in Guangzhou. Following doctrines of the old media policy that media have an obligation to report positive news (e.g. progress the government has made, the improvement of people's living standard, etc.) rather than negative news, the Guangzhou government did not realize how serious SARS was and tried to block

---

44  It is a serious disease called severe acute respiratory syndrome (SARS). SARS first broke out in Guangdong and spread to other regions.

the news. Meanwhile, the mass media, especially the press and television that have more influence on the masses chose to follow the Guangzhou government's attitude and not publish the news. As a result, the masses had to get the news from other channels.

History contains sufficient evidence to prove that unexpected events, for instance, catastrophe, revolutions, social disturbances, conflict, etc., will make people more aware of the mass media than in other times. The masses expect more information from the media to reduce the feeling of ambiguity, and they try to gather more details from the mass media but most of the information is incomplete and fragmented (Ball-Rokeach & DeFleur, 1976).

As the independence model of mass media states, ambiguity is the main character for insufficient and conflicting information. When emergencies or other social events and conflict happen, individuals cannot apply old values or strategies to cope with new situations, and that is when the ambiguity occurs. As a result, individuals lack enough information to understand the conflict and do not know how to solve it (Ball-Rokeach & DeFleur, 1976).

Controlling the mass media means controlling how people resolve their ambiguity. The ambiguity is especially pervasive during a transformative period or unexpected events, because of the traditional values, customs and beliefs cannot provide enough information to deal with the situation, and hence, ambiguity is acute during these times. In a society junction, for instance, the transformation from an agricultural society to an industrial society, from the industrial society to the information society or a social conflict period, people become heavily dependent on the media to get more information.

The attitude formation is affected by the media and can lead to serious consequences. The news and TV drama tend to describe our surrounding environments like violence-ridden jungles, which may increase people's fear and anxiety. The insecure feeling of audiences increases during the social disorder time, and it does not matter whether they live in a horrible environment or not, but how the media describe the situation. Correspondingly, the audiences might be emotionally triggered to respond violently to other's actions because the mass media always uses the sensational news to attract their audiences. If the contents of mass media are mostly violent and sensational news, the atmosphere of the society will be very tense, and individuals become sensitive and react violently (Ball-Rokeach & DeFleur, 1976).

In order to keep the masses calm, the media chose to keep silent about the SARS disease. At the beginning of 2003, the virus started to spread very fast and got out of control. The masses got nervous and rumors were everywhere, at this time, the inhabitants in Guangzhou city started to escape the city and bought medicine such as Banlangen[45] (板蓝根), white vinegar and masks to protect themselves, as a result, the price of these goods were extremely expensive. The press, such as Yangcheng Evening News (羊城晚报), was the first to report on the unknown virus and their Internet website also tried to publish the news, however, giving in to pressure from the Guangdong government, the news was deleted. Afterwards, an SMS that wrote, "a serious unknown flu virus invades Guangzhou" was sent 126 million times within three days via mobile phones (Xia & Ye, 2003).

As SARS was a very new virus, no one knew what it was and where it came from. Meanwhile, the Guangzhou government still kept its silence. However, the unofficial press and the Internet started to publish news about SARS, this kind of media not being as strictly controlled by the government. The tabloid newspapers and Internet news outlets increased quickly during this time of social disorder, but compared to the giant news agencies their news was fragmented and unsystematic. Lacking an official resource, the masses felt lost and started to transfer the information they had about SARS by oral communication. They did not know whether the information that they transferred was right or wrong, but the truth of the information was not important, what was important was to spread the news to the most people. Meanwhile, the official mass media still kept silent, whereas tabloid news grew fast and rumors were everywhere and less trustful (Jin, 2003).

Since Guangzhou was at the edge of collapse, the government had no choice but to admit the SARS virus was very dangerous and hope the masses would calm down, not spread further rumors, and allow the government to settle the virus issue as soon as possible. Afterward, the mass media knew the attitude of the government, which means they knew the basic and bottom line when they reported the news. Finally, the main newspapers in Guangzhou city started to report on the virus and society was not as anxious and nervous as before; the situation became less tense (Xia & Ye, 2003).

As SARS is an important case to study social crisis, the spread of information during this chaotic period can be analyzed to suggest how the media, masses and the government react. An investigation by Guangzhou University, using questionnaires about how people got SARS information and how their attitudes were formed, found that oral communication was the main method of information dissemination in this instance. The Guangzhou government's attempt to silence the SARS story led to the public believing rumors and spreading misinformation (Xu, 2005).

---

45 Traditional Chinese medicine made by some herbs.

Another investigation completed by Nanjing University suggested that individuals with a good educational background, especially those with a graduate degree, could get SARS related news earlier than other groups. In addition, high wage earners got the information on SARS earlier than the lower salary groups (Du, 2003). This indicates that the "information rich" can get news earlier than the "information poor" in the information society. However, in the case of SARS, high educational background and high income did not help individuals judge the validity of rumors; on the contrary, they were the group who promoted the spread of the rumor. There is an old traditional Chinese saying that "rumor is stopped by wise people", but during a period of social disorder, the (mis)information may confuse educated people as well.

Results were swift when the government confirmed the SARS virus was spreading in Guangzhou. Nevertheless, as the central government has higher authority compared to the local government, if it was not officially confirmed by Beijing, the masses still felt uncertain. After the legitimacy of the virus was confirmed by Beijing, the masses were less anxious and the whole society started to return to order and took measures against the spread of SARS. From the beginning when the virus started in November until it was first exposed by the press in early January, and then to confirmation by the Guangzhou government and the central government, took nearly four months. It is worth to note that before Beijing legitimized the "public emergency" that was happening in China, other cities and provinces attached little importance to SARS and believed the SARS in Guangdong would not spread. On 20th April 2003, the Beijing government held a meeting and announced that two leaders were fired because of their mistakes, which invoked a public uproar within the whole country. Since then, the SARS virus was officially confirmed as the deadly virus that jeopardized public health (Wu, 2014).

In this information society, by analyzing the case of the SARS virus, it can be seen that rumors are not stopped by wise people, but stopped by information transparency. The central government realized that it was crucial to release the current information about SARS to make the society less anxious. Afterwards, almost every main newspaper, television program, Internet website and radio were filled with news of SARS. However, the fragmental information did not help to decrease the anxiety of the masses, since facts and rumors mixed together, which increased the difficulty to find the truth. As Segal's law suggests that a person with one watch knows what time it is, but a person with two watches is never certain of the time. The SARS news experience of going from hidden to being exposed under the sun implies the potential pitfalls of having too much potentially conflicting and random information (Bloch, 2003). The sudden overload of news and information made it difficult for the masses to make an informed judgment on the situation.

In 2003, the Internet in China was just at the threshold and the government was not prepared to enter an information society and lacked experience in how to deal with a public emergency. When at last the governmental media introduced the virus, they did systematically introduce details of SARS and updated the information timely. Finally, the situation got under control, and rumors had less space to transmit as the information came from official sources and began to dominate mass media. As mentioned earlier concerning group psychology, when most of the people in a group believe the same information, then the group will believe the common majority-held information and individuals in the group will be less likely to form their own personal values.

Overall, the SARS outbreak was a chance to observe the performance of mass media and the reaction of the government and the masses during a social emergency and period of disorder. When Ulrich Beck defined the risk society, he pointed out how fragile our contemporary society is (Beck, 1992). In terms of SARS, within only five days from the first SMS transmission among mobile phone users, Guangzhou City was led to the edge of panic: the soaring price of goods, mass anxiety, the lost government, and the silent mass media, etc.

After SARS, the government learned that when a social crisis happens, the media should report news timely and systematically explain and update the development of the crisis. Nowadays, when surfing the Chinese government websites, it can be found that the official press and websites prefer dedicating a huge page just to focus on one issue. It is common now that the official media and mass media always systematically introduce one issue or event with a holistic view, which is useful for the government to win support from the masses and the masses are aware of the situation.

## Guo Meimei

Social change is not triggered by the single change of one subsystem, but the change of one subsystem will always be invoked by a series of changes. The policy system contains many subsystems, for example, the one-child policy is one of them. Similarly, there are many subsystems of the communication system as well, weibo (微博), it is a Chinese microblog, like Twitter) is an important branch of the subsystem of the communication subsystem.[46] It is worth noting that the

---

46 In September 2009, the Sina Weibo (新浪微博) enters into the public's eyes. Similar to Twitter and other mircoblog platforms, it provides a virtual platform for users to

one-child policy is closely connected to the development of weibo. According to statistics from the mass media research laboratory of Shanghai Jiao Tong University, in 2010, at the beginning of weibo's diffusion, there were 72 hot topics which were widely discussed, among these, 22 were exposed by weibo. Weibo functions as an information production factory and serves as a public discussion arena (Yao & An, 2012).

On 21st June 2011, a girl on the Internet called "Guo Meimei Baby" aroused a trust crisis concerning the Red Cross of China. She uploaded many pictures about her luxury cars, bags, villa, etc. Within two hours, this weibo was transmitted by thousands of weibo users. Her pretty face, luxurious life together with her young age (20 years old) invoked the curiosity of the masses. Only three days after she exposed her private life, 70,0000 netizens started to pay attention to this girl, two months later the followers of her weibo increased to 580,000. The masses, especially the netizens, started to search for who exactly is Guo Meimei, and they found she had regarded herself as the Commercial Manager of the Red Cross Society of China. Afterwards, angry emotions brought on by the resentment of the rich, perceived cheating and inequality pointed to this girl (Qian, 2012). She detonated a bomb that hid under the peaceful surface of society and a massive online hunt started to investigate this girl. Many questions were waiting for the netizens such as who is Guo Meimei, why this young girl has so much money; moreover, what is the relationship between her and the Red Cross?

As more attention focused on this girl, the press and television also joined the effort to search for Guo Meimei. The Internet provided an open and highly interactive public sphere, and since so many resources of news and information could be explored, truth and rumors intertwined with each other. Someone said she was the daughter of the CEO of the Red Cross as their family name is the same. Someone else said she might be having affairs with some rich people who work in the Red Cross. The netizens were like Internet polices who were keen to investigate the real identity of Guo Meimei. The massive online search had pushed the Red Cross to clarify they did not have any relationship with Guo

---

interact with the outside. The news on weibo update very fast, billions of users via weibo shuttle between the role of information senders and readers. The masses regard 2010 as the first year of the weibo era, as many giant Internet companies launched their weibo products (Sina, Tecent, NetEase, etc). Different companies competed with each other to win the markets, which invoked the chaotic Internet microblog market as well. When the Internet was introduced to China, it changed the pattern of information and knowledge production, however, the wide spread of weibo brought a radical reform of the intercommunication of information and knowledge.

Meimei, but it seems the masses were incredulous and not satisfied with the answer, as inquiry was not stopped by the declaration. On 4th July 2011, Guo Tao the CEO from a company called Zhong Hong Bo Ai (中红博爱), said that Guo Meimei is the lover of Wang Jun, the former CEO of his company (Hao & Lu, 2012).

During the entire online hunting procedure, the police and official organs kept restively silent compared to the massive online hunting of the netizens. The netizens acted like soldiers defending morals and ethics, and continued searching and investigating the whole matter. The Beijing Police Station tried to clarify that the Red Cross did not have a relationship with Guo Meimei and that Guo Meimei was actually the lover of the former CEO of the Zhong Hong Bo Ai limited company, but it seemed the masses were not satisfied with the results, they felt cheated and their anger was transferred from Guo Meimei to the Red Cross. Since Guo Meimei had become an Internet issue, the Red Cross kept silent until it finally had to hold a presentation due to the pressure from the masses (Zhao, 2012).

From the Green Dam Youth Escort (a content-control software on all new PCs) plan to the case of Guo Meimei, we can observe the change in the reaction of the official administration. Before China entered into the information society and new media age, the government controlled most of the information, and was the sole source of published news; however, the wide diffusion of ICT has shaken the situation of how and what information can be collected and disseminated by the government and media. The Red Cross wanted to keep silent to maintain its priority as it represents the administration, and they did not want to admit anything because they believed that would lead to a loss of their authority. However, it may have worked in the past but not in the information society, as the public can easily take fast action. Two weeks later, the Red Cross had to register an ID in Sina Weibo for the sake of participating in the debate, so the advantage of cyber-democracy began to show.

In the case of Guo Meimei, the masses were angry not about her rich life, but her relationship with the Red Cross. Her weibo ID as the Business Manager of the Red Cross of China made the masses feel cheated. During the Wenchuan earthquake, the Red Cross received impressive donations from individuals and companies, albeit the mass media sometimes exposed that the donations did not reach the victims of the earthquake, still, the masses believed the Red Cross had nothing to do with this consequence, the main problem was the local governments did not use the donations properly.

After Guo Meimei had become a hot topic, the masses were outraged at the Red Cross. According to the Red Cross of Shenzhen, after the case of Guo

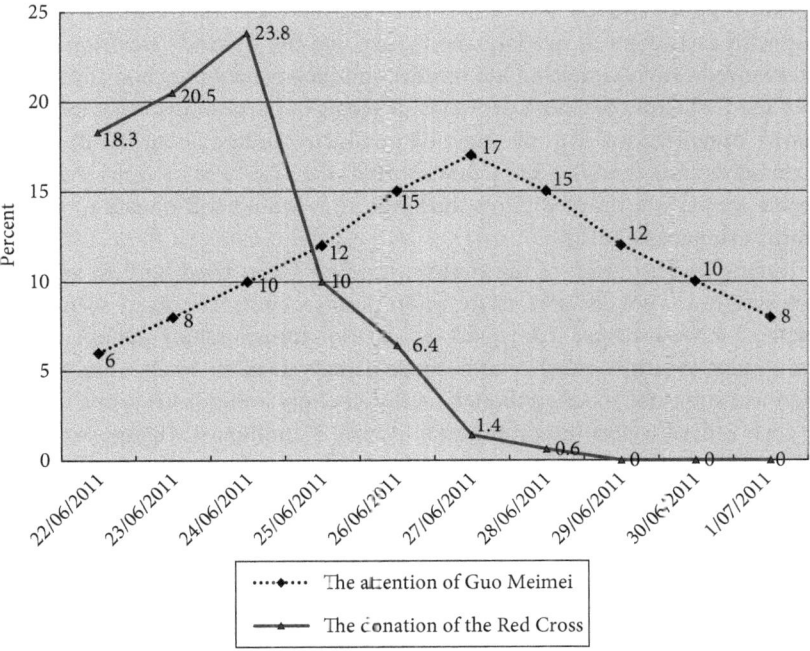

**Fig. 5-1:** The relationship between the rate of public attention to Guo Meimei and the donation amount

Source: Gui and Xu (2012).

Meimei, they only received one hundred donation items and until 3rd August 2010, the Fushan Red Cross did not receive any donations. Zhao Baige, the new leader of the Red Cross said that the individual donations decreased dramatically, the trust crisis brought on by Guo Meimei was a tragedy for the Red Cross of China, and was regarded as the "three days destroy hundred years' endeavor" (Yao & An, 2012).

In 2011, after Guo Meimei had received so much attention from the masses, the donation of charitable goods from June to August decreased by 86.6 % in 2012 (Gui & Xu, 2012). Fig. 5-1 shows the rate of public attention to Guo Meimei and the donation amount from 22nd June to 1st July 2011, indicating that donations had declined rapidly since the public attention on Guo Meimei increased. It also shows that the decrease in attention to Guo Meimei did not get the donation rate to return to the regular pattern, and in a short time, it was irreversible.

Guo Meimei and the Red Cross both clarified that they had no relationship with each other in official mass media, and the Central Television Station interviewed Guo Meimei and her mother and they denied they had any relation with the Red Cross of China. However, it seems the masses lost trust in China's charity organizations. An investigation conducted online about the likelihood of donating money to the Red Cross showed that 77.6 % of netizens would not donate money via the Red Cross, however, they would still donate to another charity (Gui & Xu, 2012).

Having lost the trust of the masses, the Red Cross tried hard to regain it. However, it was not effective. In the information society, the society seems more fragile than we anticipate. The quick information transmissions and open online discussions are more effective at homogenizing values. In 2013, when Sichuan Ya'an was subjected to an earthquake, the Red Cross immodestly sent a message on their official weibo and promised to donate 51 million RMB (approximately seven million dollars) to the earthquake struck areas (Li, 2014). However, the strategy the Red Cross used to win the masses turned out in vain, given that comments about this weibo were occupied by criticism, which implied that the masses would rather deliver materials or send money directly to the earthquake-hit area rather than donate to the Red Cross.

Why did Guo Meimei invoke a trust crisis of the Red Cross? The main reason can be attributed to the Red Cross's non-transparency to the masses. It is a secret organization, which does not belong to the government, but donations they get are distributed by the government. Guo Meimei was just a detonator, the problem of the Red Cross itself and its operation strategy were the main reasons the issue developed into a trust crisis.

Secondly, the Red Cross reacted improperly after the trust crisis. 21.1 % of netizens claimed that the Red Cross used an old way to deal with this crisis, as its bureaucracy organization system could not adapt to the information society (Gui & Xu, 2012). The Red Cross tried to keep up its prestige instead of resolving the crisis as soon as possible. As a result, it missed its chance to solve the crisis. The negative news about the Red Cross such as "expensive tent", "huge administration fees" and so on blossomed online; these kinds of news was sometimes fake news which wanted to collect website clicks, however, the news hurt the reputation of the Red Cross and the Red Cross lagged to clarify and take action.

Thirdly, in the information age, fake news and rumors always intertwine together and it is difficult for the masses to distinguish. Therefore, fake news and the virtual Internet environment make the masses skeptical about all news (Gui & Xu, 2012).

The Red Cross of China took a series of reforms. For instance, the information of donations and distributions became accessible for everybody, they released the fiscal budget and expenditure regularly, and increased the transparency of the administration and so on. The new leader of the Red Cross, Zhao Baige had promised that after three years, if the Red Cross had still lost the trust of the masses, she would leave. However, after three years, what were the results? The masses still had not forgiven the Red Cross, even though the Red Cross's involvement in this Internet wave made no sense as they were also a victim. Guo Meimei and Wang Jun, the former CEO of Zhong Hong Bo Ai were sent into prison, the leader Zhao Baige kept her promise and resigned from the Red Cross.

In the information society, the government and the masses both interact and compromise in the Internet arena, but it is good to see that the Internet has provided a platform that has dramatically changed the social system Since the Guo Meimei Internet event, many government organs registered their official ID on the platform of Sina Weibo in order to communicate with the masses. However, the influence of media cannot be underestimated, since it is striking that one Internet popular person destroyed the whole charity field. The future of charity in China is not positive, the anger of the masses needs a long time to release, as the memory of the society is easier to be formed than fade away.

## Conclusion

Information is fulfilled by digital devices and thus individuals get fragmented sensational news every day. The masses cannot get the right information to judge which part is right, as all the information they get is from the mass media. At the beginning, the masses might look forward to the update of news every day, so they are more aware and engaged in the social issues in the short term, but in the long-term, they might lose interest in this fake news and their social trust will decrease as well. The sensational news posted on the Internet increased the individuals' feeling of insecurity and the media content in China transformed from positive news to mainly negative news, which resulted in a trust crisis.

The emergence of the Internet might add difficulty, as in the information age, because the so-called new media allows users to not only passively receive information, they can also be transmitters, publishers and producers. The trust crisis which was triggered by the fast diffusion of ICT demonstrates that the relationship between government, media, and the masses have entered a new age.

# 6 How an Indigenous Society Develops in the Wave of Modernization

Tall mountains limit the connection between Guizhou and other parts of China, and many regions are still unconnected due to the geographical difficulties. In the wave of modernization, local governments have utilized historical buildings and the local minority's culture to develop the tourism economy. When the local governments started to develop a tourism economy, this outside force collided with the indigenous culture and a vivid picture developed, which can be used to deduce the interlinked relationship between modernization and tradition, and the role of modern mass media.

In this part, I will use the villages I investigated as a window, with the sociological approach that I regarded as a magnifying glass, aiming to draw a map which follows a line of ICT to see how an indigenous society transforms into a relatively modern society. The purpose of choosing to investigate the minority places was because the tourism economy is like an outside force that opened the door of the local village, allowing entry of the fast modernization process.

## A Window to Observe

In the southwest part of China, there are many minority regions, which are geographically difficult to reach. Guizhou province is one of the provinces with a low GDP and per capita income. The tourism in Guizhou started between 1995 to 2000 because of the beautiful landscape and various minority related sites and cultural events. The indigenous culture attracts many tourists that support a tourism economy, though the rushing tourists and the local people returning from the large cities break the peaceful life of the local villages and accelerates the process of modernization.

These places provide good examples to study how a traditional area transforms into the modern society and how indigenous culture is influenced by modern technologies. The indigenous environment is helpful to understand: To what degree does the modernization change in traditional societies? In which way do the modern values shape the traditional values and individuals' behavior? What is the role of mass media in the process of modernization?

The investigations began in 2010 with a series of qualitative interviews and quantitative questionnaires in the sample places. The first investigation took place from June 2010 to September 2010 in Longjiao village, Wanggang village, and Xijiang village and collected 954 questionnaires and 75 interviews. The second investigation started from July 2011 to August 2011, collecting 1048 questionnaires and 143 interviews in Pianpo, Xijiang, Langde, and Biasha. The third investigation collected 85 questionnaires and 20 interviews with the help of one assistant in Xijiang village from August to October in 2015.

The investigated region is a tourist site. According to the records of local governments, Pianpo started in 2010, with the leader called Yang playing an important role since the beginning. Zhenshan started in 1995, Xijiang in 1998, Langde in 1997 and Biasha began in 1999. All of these places are government-oriented tourism economies. Xijiang is the most successful tourism economy among the survey sites.

Why I regard them as a traditional area involved in the process of modernization? First, it is necessary to shortly explain the difficulties of research in the rural area. It is common when doing an investigation in China, and not only in the minority regions, that many conflicts between villagers and local governments will arise due to the lack of trust in modern Chinese society. Consequently, when undertaking an investigation in China, one is always faced with many obstacles. The investigators and researchers have to overcome the fear of the masses and also the intervention of local governments.

## The Difficulty of Entering into the Local Community

These areas are remote and can only be reached by car. The local governments want to utilize their traditional culture to develop the economy. Because of the transportation obstacles, these places are lagging behind in the process of modernization compared to other regions. Currently, the local government that caters to the tourism industry is considering building a high-speed train station to attract more tourists.

It is not easy for researchers to enter the minority places specifically to do field investigations. The traditional culture and modern history teach the masses that they should very carefully express their opinions. Since the conflict between local government and the masses has become acuter over various issues, the local governments are very sensitive to any kind of investigations and journalistic reports.

The first field research is a project supported by the National Natural Science Foundation of China. We entered the region as a group of eight investigators. Before entering, we asked an officer who works in the Guizhou province to call local officers to ask that they support our investigation. However, even though the officer who works in the provincial government had called the local government, the transfer of information was not timely received and not all the local officers got the notification, which in turn our investigation was stopped two times. One time, the local officer drove his motorbike in front of us and questioned about what did we do, why did we collect the questionnaires and where did we come from, etc. Afterward, he seized several questionnaires that were just finished.

For a massive investigation in China, without the support of local government, it is very difficult for researchers to do an investigation independently. It is not only because the local governments worry about researchers and journalists exposing conflict, frauds, or corruption in the region, but also because the local people fear that if they say something wrong there might be bad consequences. The traditional culture teaches the masses that it is always a risk to talk about politics. When the Cultural Revolution occurred, this value was intensified and the dark social memory (Cultural Revolution, Tiananmen Square movement, etc.) still looms over the masses.

## Background of the Samples

The sample places are located in Guizhou province, Qiandongnan Miao, and Dong Autonomous Prefecture. It is an autonomous prefecture in the southeast part of China (see Fig. 6-1). The investigation fields are in a mountainous area; before the tourism economy, agriculture was the main industry in this region.

### *Xijiang*

Xijiang Miaozhai village (西江苗寨) is located in the northeast of Qiandongnan Miao and Dong Autonomous Prefecture. It is about 36 kilometers from Leishan County (雷山县) and about 35 kilometers away from Kaili (凯里市). In the Miao language, Xijiang is called 'dlib Jang', before 1729 it was called 'Xian Xiang' 仙祥. Xian, meaning immortal, and xiang, meaning peaceful harmony were translated according to the pronunciation of the Miao dialect. From 1729 to 1961, Xijiang was called "Ji Jiang". In 1961, it changed to Xijiang and that is the name used now (Yang, 1997). Xijiang Miaozhai includes eight small villages, namely Yangpai, Dongyin, Pingzhai, Yetong, Yedong, Nangui, Wuga, and Yehao. Among the eight villages, only Yaangpai, Dongyin, Pingzhai, and Nangui of Xijiang were investigated.

The Miao (苗; Pinyin: Miáo) are an ethnic group which are mainly located in south China and southeast Asia. They are recognized by the government of China as one of the 55 official minority groups. According to the Sixth National Population Census of the People's Republic of China, there are 9,426,007 Miao people. It is the fifth biggest minority in China, accounting for 0.7 % of China's

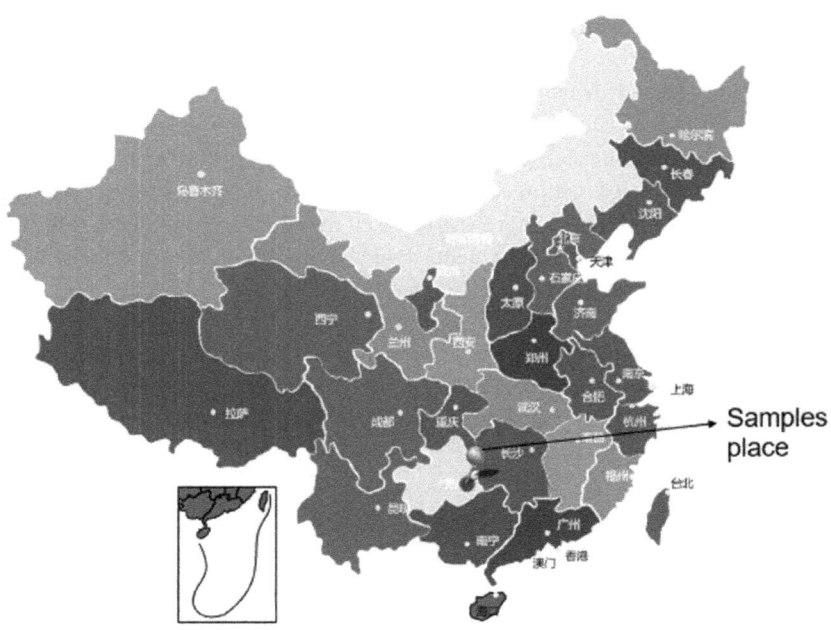

**Fig. 6-1:** Location of the sampling area.
Source: Author's compilation.

population. In terms of Xijiang, most of the people (99.5 %) living in Xijiang are Miao minority people—so it is also known as Xijiang thousand Miao villages. The Xijiang government statistics show it has 1086 households and 5287 Miao people.[47]

Xijiang is regarded as the most original Miao culture in the world, and it is the biggest Miao village in the world. The inhabitants of Miao people are mostly in the mountain area in the southwest part of China (Schein, 2000). According to the investigation, before the tourism economy, agriculture is the main resource contributing to the local people's revenue. The geography and dialect unique to Xijiang have protected their traditional culture. With the trend of globalization, together with their booming tourism economy, the local people have realized that they can utilize their indigenous culture to attract tourists and earn money.

---

47 The statistic is collected in the Xijiang government during the field investigation in 2010 and 2011.

## Zhaoxing Dong Village

Zhaoxing Dong Village (肇兴侗寨) is the biggest and oldest village of Kam people. Zhaoxing Dong Village is located in southeastern Guizhou, which belongs to Liping county of the Qiandongnan Miao and Dong autonomous prefecture. Kam people are the 12$^{th}$ biggest minority in China. According to the Sixth National Population Census of the People's Republic of China, the Kam population is 2,870,000 (Luo, 2004).

Similar to Xijiang, most of the inhabitants in Zhaoxing are Kam minority and located in mountain basins. More than 98 % of the people in Zhaoxin are Kam people. The field research investigated 9 villages: Zhaoxing, Zhaoxing Shangzhai, Zhaoxing Zhongzhai, Yilun, Xiage Shangzhai, Tangan, Jitang, Jitang Shangzhai, and Dengjiang. The villages contain 920 households and 3496 inhabitants. It is important to address that most of the inhabitants' family name is Lu (陆), inside of the Lu family there are five groups, namely *ren* (仁) group, *yi* (义) group, *li* (理) group, *zhi* (智) group and *xin* (信) group. The name of these five groups can be traced to *wuchang* (五常) from Confucianism: *ren* means people who are friendly and harmony; *yi* means people who have the will to help others; *li* means people who have morals and good behavior; *zhi* means people who have wisdom; *xin* means people who can trust other people.[48]

Zhaoxing Dong Village is one of the six most beautiful villages in China. It was also awarded the top thirty-three attractive tourist places in the world by *National Geographic and National Geographic Travel*.[49] The traditional architecture *diaojiao lou* (吊脚楼) and the drum tower are famous for their complex design which are ranked in the list of Guinness Book of World Records, and the Dong dance and Kam Dong choirs are in the Representative List of the Intangible Cultural Heritage of Humanity.

## Longjiao Village

Longjiao village (陇脚村) is located in Xiangzhigou scenery park 42 km from Guiyang City, the main capital of the Guizhou province. It is also a minority village, most of the people living there are Bouyei (百越) people, which are the 11$^{th}$ largest of the 56 ethnic groups officially recognized by the People's Republic of China. Since Longjiao village is near Guiyang city, the local government utilizes

---

48  The information was collected in the local government office.
49  See http://www.discoverchinatours.com/travel-guide/zhaoxing/zhaoxing-dong-village/. Accessed November 2, 2015.

the advantages of their traditional culture to attract tourists from Guiyang. The target customers are people who live in the big cities, especially Guiyang people who come for recreation in Longjiao Village.

Before the tourism economy, the main revenue of Longjiao village was agriculture. With the modernization as well as the pressure to increase GDP from the top of government, the local government realized that organic agriculture and indigenous culture are treasures to boom the economy. This region is where the ancestors of the Bouyei were the first to cultivate rice; and now they are successfully combining the local property with the trend of modernization.

### Wanggang Village

Wanggang Village (王岗村) is located in Guizhou province 38 km from Guiyang City. It has 202 households and 849 people, among them, 99 % are Bouyei (百越) people. The village has six groups, namely Wanggang Liuzu, Gizu, Shangjie, Xiajie, Shanghewan, and Xiahewan.[50] Similar to Longjiao Village, Wanggang village also uses their traditional culture and beautiful landscape to develop their economy. By renewing the traditional architecture, and restoring the original layout, the village has deeply explored their traditional culture to find a way to survive the ongoing modernization.[51]

## A Traditional Society

I have mentioned that the investigated region still keep their traditional culture, but what is their traditional indigenous culture? Brown (1969) proposed that there are four aspects to study traditional society and modern society: economy, social structure, political institution, and values.

Among the four aspects, it is worth to note the last dimension—values. Max Weber proposed "the spirit of capitalism", which raises the question of what is the sprite of modernity? As modernization is a multidimensional process, one of the important dimensions is the attitude of individuals, and the attitude developing man has experienced from the traditional to the contemporary era. When social scientists study this, the usual approach is under the context of urbanization and industrialization, where an emphasis is on whether a rural person who moves from village to city can participate in the modern industry and become a "modern industrial man" (Horowitz, 1976, p. 263).

---

50  The information of the village is based on the investigation of the local officers.
51  The information was collected in the local government office.

Before the tourism economy began, the local villagers, especially young men had already experienced work in the big cities and attempted to change themselves into a modern industrial man. According to the local people's memory, since the end of the 1980s and beginning of the 1990s, young villagers started to leave the village to find work, namely *dagong* 打工 (打 means work and process, 工 means jobs). Before traveling out for *dagong*, the villages did not have communication with the outside except for the occasional foreigner stopping by on a travel or to do research.

In order to know the changes, it is important to find their indigenous culture. After checking some documents from the local government, and asking local people during the investigation, I will proceed from the four aspects as Brown proposed, namely economy, social structure, political institution and values to introduce their indigenous culture and why I regard the place as a traditional region involved in a modernization process.

## Economy: Agricultural Society with Pervasive ICT Diffusion

According to the reports from Qiandongnan Miao and Dong Autonomous Prefecture for the last five years[52] (from 2011 to 2015), agriculture is the main revenue of the local people. Most of the local people are involved in the basic production, such as agriculture, forestry, and animal husbandry. Based on the data from Economic and Social Development of Qiandongnan Miao and Dong Autonomous Prefecture (ESDQMDAP), Fig. 6-2 suggests that since 2011, the tertiary economic sector has increased faster than the secondary and people have started to transfer from the primary to the secondary and tertiary sectors.

The second and third investigations suggest that agriculture is the main revenue resource of the local villages. During the first and second investigations, the investigators were asked to enter into local people's houses and observe their living standard. The results suggest that the living conditions of the local villagers were simple. The average yearly revenue of one household was about 1500 dollars. The basic house appliances are televisions, phones, refrigerators, and washing machines. Except for Xijiang's high school, the other investigated villages have only primary schools and teachers have to teach many subjects. A teacher called Hou in Longjiao village said his English is bad, but he has to teach English, and at the same time, he has to teach Chinese as well, because there is no teacher. Local villagers who are rich will send children to study outside, but they are just a few cases. Most of the villagers do not have a good education. The people who

---

52 See http://www.qdn.gov.cn/xxgk/zdgk/tjxx/tjnb/. Accessed August 18, 2016.

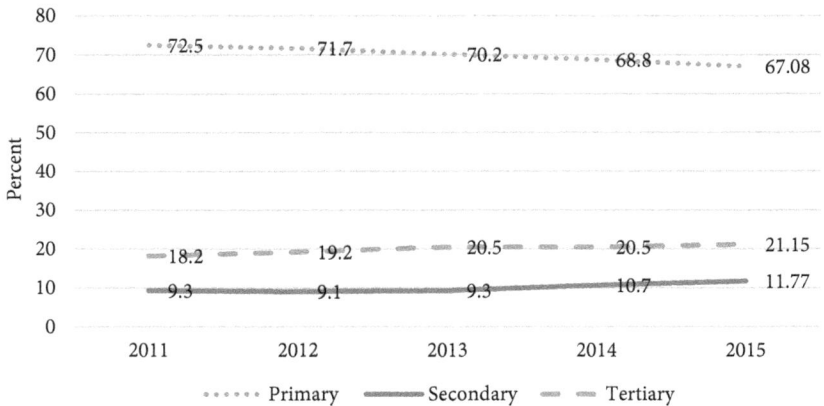

Fig. 6-2: Population distribution of economic sectors of the Qiandongnan Miao and Dong Autonomous Prefecture.
Source: Economic and Social Development of Qiandongnan Miao and Dong Autonomous Prefecture (ESDQMDAP, 2016).

are well-educated prefer to work in big cities and the developed region, nobody wants to work here if they are a university graduate. Before the villages started to develop their tourism economy, many people had already left their hometowns and migrated to the east coast cities to work. The people who remain are called the "38, 61, 99 army" (三八，六一，九九部队),[53] which is an ironic way to describe only the middle-aged women, children and elderly people left behind. For young people and adult men, the village was boring and without an opportunity to earn money.

Tab. 6-1 shows an interesting phenomenon: from the third investigation in 2015, which was only undertaken in Xijiang and limited to local Xijiang villagers, is that there are less people working on crop farming and tourism services. It looks contradictory that as the most successful tourist region, there are less local people working on tourism services and individually-owned businesses than in other regions. The reason is the tourism services are mostly run by outsiders, as well as tourist agencies, tourist companies and so on. The shops near the main

---

53  38, Pinyin sanba 三八, comes from the International Women's Day (IWD) which is on March 8, it points to women. 61, Pinyin liuyi 六一, comes from the Children's Day which is on June 1, here means Children. 99, Pinyin jiushijiu 九十九, here means elderly people.

**Tab. 6-1:** Revenue resources of the investigated region in 2011 and 2015

| Revenue resource | 2011 (Percentage) | 2015 (Percentage) |
|---|---|---|
| Crop farming | 56.2 | 25.0 |
| Animal husbandry | 14.1 | 10.7 |
| Dagong (work outside) | 37.0 | 22.6 |
| Individually-owned business | 21.9 | 20.2 |
| Tourism service | 19.0 | 11.9 |
| Household sideline production | 5.6 | 19.0 |
| Other | 8.4 | 4.8 |
| Total | 162.3 (1665) | 114.2 (96) |

Source: Author's calculation.
Note: The question is a multiple-question. In 2011, the investigated area were Pianpo, Xijiang, Langde, and Biasha. In 2015, the investigated region was Xijiang.

street of Xijiang are mostly rented by outsiders whereas local people who live in the mountains have less chance to participate in the tourism services. The local villages and outsiders sometimes have conflicts and most local villagers think the outsiders are destroying their local culture and occupying their businesses.[54]

## Social Structure

Regarding social structure, the difference between the traditional and modern society is that in the traditional society the dominant social unit is family and interactions between individuals are face to face, whereas the modern society is more complex, bureaucratized, and highly differentiated. In order to find changes in social structure, I will examine three dimensions: entertainment, family unit and the roles of the elderly.

The reason to choose the three dimensions is firstly, entertainment is an important factor to observe how individuals interact with each other and the social network of local villagers Secondly, the family as an essential unit in Chinese society has a long history and has witnessed many reforms. The investigated region has a strong family or clan structure in the past, and within the indigenous culture, rituals are highly developed. Thirdly, the elderly people in the villages were traditionally regarded as an authority group in the region.

---

54 I will return to this whole general question at great length. Outsiders who enter into the village are better suited to combine modern values to cater to tourists, whereas the local villagers benefit less.

However, modern technology has established a new rule in the society and has broken the old social orders. With the influence of a tourism economy, workers from the outside introduce the new culture and the impacts of modern technology, changing the role of elderly people.

Before the tourism economy, Miao people liked to chat in the "wind rain bridge" (*fengyuqiao*风雨桥), which is a traditional public sphere for local people. Kam people like to go to the *gulou* 鼓楼, a public place which is not only for their recreation, but also has a judgment function. Local villagers go to *fengyuqiao* (风雨桥, wind and rain bridge, it is a bridge for people sit and chat) and *gulou* spontaneously, and they do not need to make an appointment like a conventional meeting point. The traditional games are animal fight gambling, such as cock, bull, horse and so on.

Fei Xiaotong (1983, 1997) claimed that the Chinese society is an acquaintance society, which means within the society, the law is less important compared to relations. The society in cities is a semi-acquaintance society, relations, law, and regulations are mixed together. In an acquaintance society, a network is very important and it is also a key point to occupy an advantaged position in the whole society (Fei, 1983; Fei, 1997). Some of the investigated villages still keep agnation. For example, in Zhaoxing all local villagers have the same family name Lu (陆). In Pianpo, many local villagers have the same family name Chen (陈). These areas are famous for their low crime rate, because in an acquaintance society, people think it is a humiliation for the whole family and they will lose face if a family member is accused of illegal activity such as theft, sexual immorality, false testimony, slander, etc.

Elderly people as a luminary group used to represent tradition and transferred traditional values. Each investigated village has several *zhailao* 寨老 (寨means village, 老means elderly and knowledgeable people). *Zhailao* have rights to judge conflicts between villagers, whether deviant people need to be punished or not and how to punish them, announce important festivals and make big decisions for the villagers and so on. Currently, the role of *zhailao* is changed but they still are a symbol of tradition.

## Political Institution

According to Brown (1969), the traditional states resemble large families, whereas modern states resemble specialized associations. Political functions are assigned to different categories of people. As mentioned earlier, the investigated region has strong agnate relations, with the villagers regarding elderly people as an authority group, which make the big decisions for the villages. The political roles of elderly people, such as miaowang (苗王, village leader), zhailao (寨老, village elder knowledgeable people), guzangtou (鼓藏头, drum festival leader), etc., are gradually being replaced by the local government so that they play roles only during their traditional festival period. The original function of the local authority was less important before the tourism economy, however, in order to stress the special political institutional system of the indigenous culture, their roles are highlighted again and many tourists visit them. The miaowang (苗王) house in Xijiang became a museum and popular sight spot. The local officers are elected by local villagers and controlled by the upper government, which is similar to other places in China.

## Values

The investigated region is less commercialized and the society is very homogeneous. The elderly people are regarded as an authority group, they wear traditional clothes, only celebrate their own festivals, still use guns for hunting, have special customs in the wedding ceremony or funeral, use their herbs for medicine and so on. The situation in these areas beginning a tourism economy is similar to when China opened the door to the outside in the early 1990s.

That nature is organized by invisible souls is deeply ingrained in the Miao people's minds, a belief system known as animism. Totemism, naturism and ancestor worship are essential values of the indigenous culture. In the past, Miao people rarely celebrated the Chinese New Year Festival (春节, also called spring festival), only celebrating Miao New Year Guzangjie (鼓藏节), which is the most important and ceremonious festival in the Miao community. Gu (鼓) means drum, it always made from a maple tree, as the Miao believe their progenitors began as a maple tree, butterfly and bird. In Biasha, the local Miao villagers have a strong worship of trees. When local villagers give birth, their family will plant a tree, when they die, the family will use the tree to make a coffin. The Kam people in Zhaoxing worship the sun and animism also influence the local values deeply. Therefore, the traditional clothes have many patterns that look like the sun and sunlight. Bouyei people worship mountains, water, earth, sun, rivers, sky, etc. The dragon dance reflects the shadow of the original totem worship of the dragon.

## The Pervasive Usage of ICT in the Minority Area

Until the end of the 1980s, Xijiang lacked a modern road connected to the outside; till the 1990s, Biasha did not have a running water system; but by the end of 2010, most of the investigated region had installed modern communication devices such as phones and televisions. What happened to these villages in the past few decades?

As tourism started, thousands of tourists rushed into the minority region and modern values and the modern lifestyle had a strong impact on local people's daily life. At the beginning of the tourism economy, the local governments awarded tourist businesses to local villagers and motivated them to show their indigenous culture.[55] At the same time, the local schools became more focused on teaching Mandarin instead of their dialects, and they started to set a course to teach their traditional skills[56] because the tourism economy made them aware of the specialness of the indigenous culture. People who worked outside came back to their hometown and brought new knowledge and modern values to the villages (according to the field investigations, more than 55 % of people think that those who go outside to work will come back). Meanwhile, the inhabitants began to reflect on their own traditions and become much more confident about the local culture.

Throughout the modernization process, the local government has played an important role to improve the tourism economy and spread technologies. Firstly, the local government followed the policy of the central government to stimulate the diffusion of modern technology.[57] They set up reimbursement policies to help villagers buy televisions, refrigerators, smartphones, washing machines and so on. Secondly, they built roads and invested a lot for infrastructure—the phone networks were established and the wireless local area networks covered whole villages, which was even earlier and more advanced than some developed areas in eastern China. Thirdly, they found investment companies to cooperate with tourism companies to develop the local culture together. Last but not the least, they mobilized the local villagers to participate in the modernization. The minority places leaped immediately into the information society, bypassing the

---

55 For example, in Pianpo the local government awarded the first eight families who would like to run a family hotel service 2000 to 3000 dollars in support.
56 Before the tourism economy, the setting of courses of the local school was standard, similar to other schools from majority places.
57 A new policy called *jiadian xiaxiang* (家电下乡, 2007), which purpose was to encourage the domestic consumer, especially the village consumer, to buy televisions, was a great success. After the diffusion of television reached saturation in cities, the villages and rural places were seen as a potentially big market. According to the *jiadian xiaxiang* policy, when village customers bought household appliances (refrigerator, washing machine, television, air conditioner, computer, etc.), they would get a discount and the government would reimburse money to them. Currently, the television penetration in China has reached 98 %, a huge success.

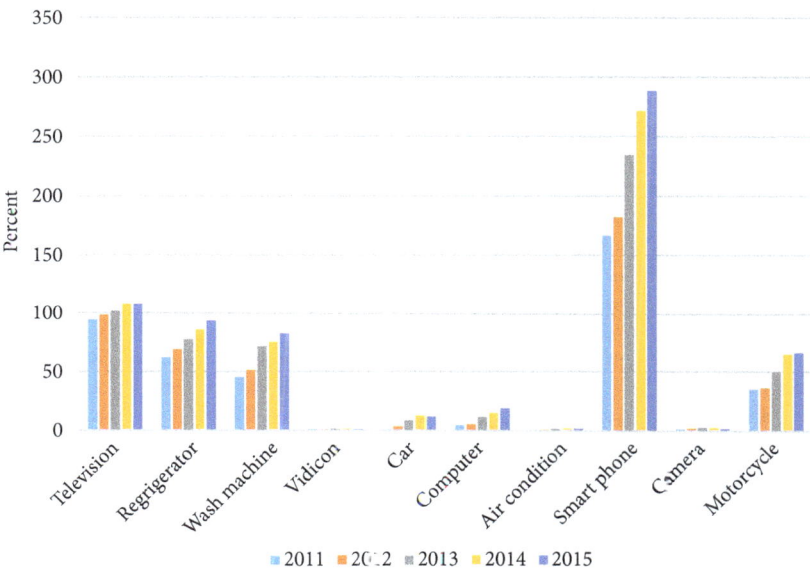

**Fig. 6-3:** The diffusion of technologies in Dong Autonomous Prefecture from 2011 to 2015.
Source: The question is a multiple-choice question. In 2011, the investigated area were Pianpo, Xijiang, Langde, and Biasha (ESDQMDAP, 2016).

industrial revolution process. As the infrastructure became largely improved, the media and tourism economy entered into the people's daily life and changed their lifestyle dramatically.

Fig. 6-3 shows that televisions and smartphones are pervasively diffused in the investigated region. The penetration of computers is lower compared to televisions, refrigerators, washing machines, smartphones, and motorcycles, however, the computer diffusion rate in 2015 increased nearly five times compared to 2011. Unlike the other appliances, computers can interact with users, transfer information and shape the values of individuals. Thus, ICT as the main technology, can stimulate social change by restructuring the communication system, and then by shaping the values of individuals, influence the traditional indigenous culture.

The diffusion of ICT in these minority areas of Guizhou province compared to other parts of China is backward. So, why is the penetration of mobile phones in the investigated region so high? Before answering that question, first, let's see what the local villagers say about the mobile phone:

"We don't need to find customers, and it is impossible to find customers. Therefore, in our village, we have a hotel committee.[58] For example, today customers come, the committee called the number 1 hotel, then the other customer will distribute to number 2, number 3…… It is very important to have a mobile phone, very convenient, especially when running a family hotel. My last mobile is broken, I bought a new one." (Yang, Pianpo, 32, Junior high school, personal interview, August 10, 2010, 1D6)

"Mobile phone helps me a lot. I cannot live without the mobile phone now. Sometimes, when I finish work in one place, my friends call me to work in another place. So I can always find work. I also do the same for my friends, if I have more work, I always share with my friends, so we can both earn money." (Luo, Pianpo, 54, architecture worker, personal interview, August 11, 2010,1D10)

"The local villagers like to go to fengyuqiao to chat and relax, before, there were a lot of people, you didn't need to make an appointment with friends, you could always find people there. Nowadays, our local government built several fengyuqiao, but there are only a few people that go to fengyuqiao to relax. We use phones to contact friends, and set a meeting point and time, for example, let's meet at a tea bar, then friends come." (Li, Xijiang, male, high school, personal interview, August 5, 2010, 2A13)

"All my friends have a mobile phone, I feel a loss of face (meimianzi 没面子) that I don't have a mobile phone. Of course, my friends won't say anything, but I feel I am less involved in some activities, I feel without a phone, I am isolated." (Hou, 13, Xijiang, personal interview, August 17, 2010, 1B10)

Thus, the field research suggests that the mobile phone gives local villagers more opportunities to earn money and makes life much easier. The local children use mobile phones very early, in 2011 more than half of the students (age: 10-17) have their own mobile phones. For the local people, without a mobile phone, they feel a loss of face. Unlike other places in China, the fixed-phone is still popular in these regions. When analyzing the media diffusion model, I pointed out that when the penetration of fixed-phones in developed countries started to decrease, the diffusion of it in developing countries, no matter the saturation level, would follow the trend. This phenomenon is proved in the whole of China but it is only partly proved in these sample places, for example, according to the third investigation, 86.9 % of households still keep their fixed-phone.

During the first and second investigations, many villagers said that they want to cancel the contracts of their fixed-phone. Because, they said, with a mobile

---

58 In Pianpo village, the local people who run family hotels set up a hotel association. To avoid the local people compete with each other and form a bad atmosphere among the local villagers, the hotel association distributes the customers equally. The family hotels are called nongjiale 农家乐 in China. 农means peasant, 家means house, 乐means entertainment. City inhabitants like to go to the family hotel to experience the fresh air, eat fresh food and relax.

phone it is more convenient to communicate with the outside. However, after three years, why did these places still keep such a high fixed-phone subscription rate? There are two reasons that could explain. First, the older generations cannot use mobile phones. The traditional custom of the local area is that elderly parents should live with their children.[59] When explaining the difference between fixed-phones and mobile phones, a local villager made a vivid comparison, she said "a *fixed-phone is like an old book, whereas a mobile phone is like an electronic dictionary. For the old book, you can read it directly, you just need to hold it in your hands and read. However, the electronic dictionary contains more information than books and it is more complex than a book*".[60] For the young generation, ICT is like a teacher. They learn Mandarin fast with ICT, and the young generation speaks Mandarin gradually in their everyday life. However, the old generation cannot understand because of the language obstacle, but as the young generation knows Mandarin and how to use *pinyin* to type, the influence of ICT will have more influence on them than the old generations. The old generations regard the mobile phone as a new technology, which is out of their capability. Most of the elderly villagers can only use a fixed-phone, whereas computers and mobile phones are only used by a few educated villagers, and they mainly use the basic functions such as taking photos or listening to music.

As for the reasons why elderly people do not use modern technologies, the questionnaires from the third investigation suggest that most of the elderly do not have an interest in learning the new technologies (34.5 %), while 16.7 % replied that they tried to teach the elderly people, but they could not learn. According to the local traditional culture, when parents get old, the youngest child will inherit the parents' property and have the obligation to take care of the parents. Therefore, in the investigated region, most of the elderly people live with their children, and most households keep their fixed-phone just for the elderly people. As long as the elderly live with the young people, they can also benefit from the modern technology.

Second is habitus, which is originally from Aristotle and developed by Pierre Bourdieu. Habitus is "the way society becomes deposited in persons in the form of lasting dispositions, or trained capacities and structured propensities to

---

59 Not always, but in most cases, the youngest child will inherit money and the house from parents, and the child will take care of the parents. Now, it starts to change as young generations migrate to urban centers and elderly people have to live alone or go to a nursing home.
60 Investigated during the first field research on August 15, 2010, record Nr.2A5.

think, feel and act in determinant ways, which then guide them" (Navarro, 2006, p. 6). It means a person's mind and emotions are a complex result of embodying social structures. Habitus is determined by one's race, gender, education, socio-economic status, profession, religion and so on. The local people who regard the fixed-phone as one of the important household appliances, for them, the fixed-phone may not be very practical, but they have used it since they were young and are familiar with it. The fixed-phone not only has a communication function, but is also a part of social memory.

In the case of the Internet, the industrial level is the most important factor influencing the diffusion curve. The third investigation in Xijiang shows that 31 % of the villagers there have a computer. Since Xijiang is the most developed village compared to the others, the other villages will have less computer penetration. As the mobile phone can also surf the Internet, the local villagers are very satisfied with the mobile phone. Only the young generations who have education above senior high school and local villagers whose family runs a hotel have computers.

## The Tourism Modernization Process

In terms of the minority area's venture into tourism, modern values and modern lifestyle have had a strong impact on the local people's daily life. As thousands of tourists rushed into the conservative region, the local government rewarded local people who run tourist businesses, which motivated local people to learn the modern lifestyle; the local school became more focused on learning Mandarin; people who worked outside came back to their hometown and brought new knowledge and modern values to the villages. The widespread of mass media keeps individuals at home rather than interacting with other villagers. The acquaintance society is changed into a stranger society, because more and more people spend time in front of the television, whereby traditional entertainment activities have fewer people to join.

Black (1975) claims that modernization includes the improvement of knowledge, political change, economic growth, social mobilization, and psychological adaptation. In the next paragraphs, I am going to use these five variables to introduce how modernity enters the traditional environment and what changes it brings.

### Improvement of Knowledge

In the past, the investigated region had difficulty attracting intellectuals to work there. Both the central government and local government had spent a lot of

**Tab. 6-2:** Comparison of information access in 2010 and 2015

| Information access | 2010 (Percentage) | 2015 (Percentage) |
|---|---|---|
| Television | 85.4 | 88.2 |
| Internet | 22.9 | 49.4 |
| Radio | 8.3 | 1.2 |
| Newspaper and magazine | 18.2 | 2.5 |
| Oral communication | 21.7 | 8.2 |

Source: Author's calculation.
Note: The question is a multiple choices question.

effort to improve the education of the local people and encourage them to learn Mandarin, but it proved that the efforts were not effective, since more and more young students dropped out of schools and chose to work in other cities. Now the tourism not only increased the economy but also opened a window for the villages to communicate with the outside. Tourism and modernization supplied enormous information to these conservative areas and changed dramatically the local people's values and activities.

According to the first investigation, more than 97.6 % of households have television sets and half of them have more than one television set. This suggests that television has become the main access for local people to get information followed by the Internet (see Tab. 5-2). In the third investigation, I divide Internet access into the smartphone and computer, and show that smartphones have surpassed the computer to become the main Internet terminal, as 43.5 % people choose a smartphone to get information whereby only 5.9 % of interviewees were with computer access.

As I have mentioned, television is the main access for people to get information, so what is the effect of television on the local villagers? According to the first field research, more than 87 % of villagers spend their spare time viewing the television, and 45.2 % villagers like to watch the news and 27.9 % people like to watch television drama. It is important to highlight that during the investigation, many interviewees said that their knowledge about law and policies are from television programs, especially *Xinwen Lianbo*,[61]

---

61 Xinwen Lianbo 新闻联播. 新闻 means news, 联播 means broadcast. It is a daily news program broadcast between 19:00-19:30 by CCTV (China Central Television). It is one of the important programs which focus on issues about economics and politics. It is like a tradition for the Chinese family to watch Xinwen Lianbo every day, making it

*CCTV12*[62] and *CCTV7*.[63] Television sets and the Internet have replaced the radio broadcast to become the main media for local people to receive information.

Educated villagers moved out of the villages to work in east coast cities and the GDP of these villages was below the average level of the whole country. However, with the diffusion of modern media, the local villagers got a chance to know the outside world and information from the media, which has dramatically shaped their daily life. The most important is the widespread usage of ICT, which has helped the local people learn Mandarin very fast, which I regard as one of the most important consequences of the modernization.

During the field research, many interviews addressed the importance of ICT. Below is a case about how local villagers learned from the television program and utilized it to protect their rights.

> "I like to watch CCTV1 and CCTV12. Before we didn't know the central government's policy and we also didn't know how to protect ourselves when our interests are jeopardized. Now, with television programs, we know how other villagers deal with the same situation and what kind of laws we can use. When the Xijiang government passed the ticket plan, many villagers disagreed with this plan. We don't know where we can send our petition and where we can go to transmit our will to the outside. The thing is we are worried that the tickets will bring fewer tourists, and another concern is if our relatives and friends visit us we have to go to ticket office, which is located far away from the village and that we would have to drive to the ticket station to pick them up. We threatened the local government that we will go to Beijing to sue them and will destroy their offices. Some of the villagers used the Internet to post what happened in Xijiang village and we also called to CCTV12. The result was the broadcast station really made a program called 'local government passed the ticket plan, local villagers disagree to this plan". In the end, under pressure of the masses and the program invoking more attention to the provincial government, the local government held meetings and communicated with the local people to solve the problem." (Hou, 26 years old, man, high school, civil servant, personal interview August 20, 2010, 2/3B1)

Based on interviews, it seems that the local villagers learn about law from television programs. The investigation found that more than 95 % of the mobile phone users have received fraudulent messages, and they learned how to distinguish

---

one of the world's most-watched programs. During the time 7:00pm -7:30pm, other broadcast stations will broadcast Xinwen Lianbo, more than one-third of programs will stop and only broadcast this program.

62   *CCTV12* is a TV channel that belongs to CCTV (China Central Television). It produces programs related to society and law.
63   CCTV7 is a TV channel that belongs to CCTV (China Central Television). It produces programs surrounding agricultural and military programs.

the information by watching the television programs. For the indigenous area, the tourism not only increased the economy but also opened a window for the villages to communicate with the outside. Tourism and modernization provide enormous information to these conservative areas and have changed dramatically the local people's values and activities. Suddenly, they have been deluged with information and knowledge, and during the learning and adapting process, their minds and values have gradually changed as well.

## Political Change

Most Asian countries are influenced by Confucianism. They believe government plays a paternalistic role in the Confucian hierarchical system, which means the government has an obligation to make decisions and individuals have a duty to follow these rules and decisions. Everybody should be dedicated to achieving the goals of the nation instead of only caring about their own targets. As Confucius said, "The ruler is the wind. People are the grass. When the wind blows, the grass has to bend (Confucius, 2012)." Both the government and individuals accept that the government is endowed with dual roles: parents and ruler. The dual roles suggest that the mass media in Eastern culture should serve as a tool for government rather than serve individuals.

In terms of the investigated region, they are mostly a government-oriented tourism economy. Some of them are government owned and some are the local governments cooperating with private companies to run the tourism business together. The villagers regard the local governments as representatives, which have an obligation to develop local villages and local culture. The local governments play multiple roles in the relation between villages and the outside.[64] They have to improve the GDP of the village, cooperate with tourist companies, administer, protect local culture and introduce modern values.

In the past, the policies were transmitted and proposed by local governments, as the local people were less educated and could not read and speak Mandarin, making newspapers unpopular among villagers. The local governments dominated the information, which endowed them with a symbol of authority similar to the central government. With the widespread of mass media, the dominant role of local government to transmit and release information was replaced by the

---

64  I will explain the role of the government in Chapter 6. The government officers in traditional Chinese values means fumu 父母guan官, 父means father, 母 means mother, 官 means officers, fumu guan suggests that the government play a role like parents for the citizens.

modern mass media, given that more than 88 % of local villagers get their information from television instead of by notification from the local government.

The first and second field investigations show that the widespread of mass media has changed the prestige of the local governments. During the interviews, I found that the local villagers trust the central government more than local government. They prefer to watch programs from the central government, such as China Central Television (CCTV). As I had introduced in an earlier chapter, each local government has their own broadcast station, but in the investigated region, the local villagers show less interest or are even against the local television programs. For example:

> "I always watch CCTV1, 4, 5, 12 (central government's channels), I don't like local TV programs. The central government's program is designed from real issues and is more just and real compared to the local channels. The local channels only report good news and never say bad news, so I don't trust local channels". (Liang, 38, male, Miao people, high school education background, Xijiang, personal interview, July 30, 2011, 2.1B12)
> 
> "The central government always releases good policies, but local government always changes these policies. Different local governments understand differently in regard to the same policy, for instance, the same policy released by the central government will be illustrated differently between my village and our neighbor village. Last year (2009), our village experienced very heavy snow, the CCTV's news said every household will get 25 dollars, but in reality, we only got seven dollars. Another example is near the river there are many houses that were established recently, we have heard that the houses were established by the central government for us for free. But now, the local government occupied them and they rent them out to earn money."[65] (Song, 50, male, Miao people, junior high school education background, Xialongjiao, personal interview, July 28, 2010, 1B3)

The situation of news and information that used to be controlled and regarded as a private resource is now changed by the spread of ICT. The explosion of information is like a flood rushing into local people's daily life. On one side, it has consolidated the central governments' authority; on the other side, it has decreased local government authority. Since the central government has no direct contact with local people, for the local people, the central government is covered by a mysterious aura. When the central government releases a regulation or policy, they use media to inform the people, and the local government has an obligation to notify the local people of the information from the government as well. During the process, the local government works as an executor. It seems the central government bakes a cake, and the local government works in the delivery

---

65 This information is not verified but it suggests that the masses do not trust the local governments.

process. The problem is the central government rarely discusses how to distribute the cake in public, consequently, the local government always plays the role of villain.

A popular saying is the central government are benefactors, provincial governments are relatives, city governments are good people, town government are bad people, and village governments are evil (Chen, 2010). Regarding the trust of local government, to trace the change, in the third investigation, I designed a question about "whether policies implemented by the local government are the same as released by the central government?" It showed that 78.6 % of local people chose "Yes", which means most people think the local governments follow the central government's policy and still trust the local governments—the result is different from the first and second investigations.

Regarding this result, it is striking that in the third investigation (2015) most of the villagers were positive toward the local government compared to the first investigation (2010). When I investigated Xijiang village, I found that the first television set arrived in this village in 1993, apparently very late compared to other parts of China. The local villagers cannot speak Mandarin, especially females and the elderly. The inhabitants of the investigated region are mostly minorities, and most of them are illiterate with a low educational background. Before the media became widely diffused in this area, they got news from the local government or during oral communication, so the local government that controlled information resources was seen as a symbol of authority. The villagers thought the local government dominated and controlled the information and that most of the good policies from the higher government were blocked. However, tourism, as one of the consequences of modernization, followed by the spread of mass media, allowed the masses timely and free access to the outside world. This direct link to the outside world and the central government allowed comparison and subsequent judgment of the local government's integrity.

As for the local government, they lost the opportunity to control information as they did before, and now in the information society need time to adjust their way of administration. In recent years, the local government opened their own microblog and information office in order to better communicate with the masses.[66] When I interviewed an officer from Xijing, he said *"the local government established an office which aims to transfer information to local people timely. In the past, we had to share the message family by family or publish it everywhere.*

---

66 In the case of Guo Meimei, I have introduced how the Red Cross of China used weibo to communicate with the masses.

Tab. 6-3: Telephone number attitude in 2010 and 2015

| Do you think the local government knows your telephone number? | 2010 (Percentage) | 2015 (Percentage) |
| --- | --- | --- |
| Yes | 13.3 | 1.2 |
| No | 31.4 | 39.3 |
| I don't know | 55.3 | 59.5 |
| Total | 100 | 100 |

Source: Author's calculation.
Note: the question is *Do you think the local government knows your phone number?*

*Nowadays with the mobile phone, we can send the information to their phone. For example, last time we got a message that there is an east coast city needs more labor force to work there, afterward, we sent this message to the local villagers via mobile phone. The modern information communication technologies make our work much easier than before."* (Hou, 39, male, civil servant, Xijing, personal interview, August 5, 2011, 2.3B6)

Based on this statement and the previous investigations, I assumed that the widespread of ICT helped the local governments disperse information more efficiently. In China, before the widespread of media, radio (loudspeaker) was the main tool to inform messages. Nowadays, phones have become the main tool for transferring information. As the local officer said they have transmitted messages to the local people timely, but what is the opinion of local villagers? In the questionnaires (2010, 2015), a question asking whether the local people know if the local government has their mobile phone number or not was designed, in order to observe opinions of local villagers.

Tab. 6-3 shows that most local people do not know whether the local government knows their phone number or not. The third investigation in 2015 suggests more villagers think the local government does not know their contact information compared to 2010. Also, it suggests that there is a gap between local government and the masses regarding the impression of whether the local government knows their phone number or not and this gap is increasing. The local governments have the numbers of the villagers, however, the local villagers do not think so.

Based on this result, more questions were designed concerning political engagement and the media. The questions are as below: (a) participate in online voting (6.0 % with Yes, 94.0 % with No), (b) Have experience of criticizing local government or report scandal of local government online (3.6 % with Yes, 96.4 % with No), (c) Have experience giving suggestions to the local government online (6.0 % with Yes, 94.6 % with No).

The result shows that the widespread of ICT has little influence on political engagement, since less than 10 % of people have used modern media to express their political opinion. This result is different from what social scientists expect, as most studies suggest that the diffusion of media will enhance democracy and political participation (Eickelman & Anderson, 2003; McLeod et al., 1996). During the investigations, the local government said they forwarded new messages using mobile phones to the local masses, but according to the investigations, only 27.4 % people gave positive answers, and 72.5 % claimed that they have never received messages from the local government.

The information once controlled by local governments is now available to the masses timely and free. The relationship between the masses and local government has changed, because local governments lost their prestige after the wide spread of ICT. The local government also uses media tools to inform villagers but not very frequently. The political engagement of the local villages has not increased by the diffusion of ICT. The masses use media to get information, preferring to trust news from the central government and the east coast press over the local news. In addition, they rarely use mass media to participate in local political activities, which is related to the loss of credibility of the local government.

## Economic Change

The government-led tourism economy has inspired the masses to run tourism businesses. As more and more tourists are drawn to the local villages, many people have returned to their hometown, changed their houses into hotels, opened shops to sell local products, established tourist companies, and are working as tourist guides to introduce their villages and cultures. In the past, when young people left their hometown, they believed they had no future in these areas; moreover, they thought the traditional culture is like a cage that confined economic development. With the tourism economy, their self-confidence increases as they reflect on and realize the special nature of their traditional culture.

As for the fast developing tourism economy, the traditional economic system is replaced by the new economic system, which is followed by imbalanced development and inequality. For the developing countries, the gap between rich and poor is one of the biggest issues in the process of modernization. For the investigated region involved in modernization, change to their indigenous society is dramatic as they face fast economic growth and agriculture is replaced by tourism, self-employment, small business and so on.

**Dual Markets**

In the investigated region, dual markets are formed. A dual market means that there are two markets, one market for tourists with expensive prices, another "discount" market only for local villagers. Xijiang village, as the most successful tourist village, has typical double markets existing at the same time. The customers for the local trades are mainly local villagers, and they call it chicken day (*jishi*, 鸡市) and rabbit day (*tushi*, 兔市) which are calculated according to the lunar calendar (Chinese traditional calendar). The price of the local market is one-third or half price of the tourist market. The dual markets are generated by the tourism economy. Before tourism, the local market was the only place for them to buy and sell goods. Most products in the tourist market come from the local village. The local market is autogenous, growing from traditional culture and are mostly street markets, which without the need for a shop allows anyone to be a seller. As many tourists rush into the minority region, prices of goods, especially goods imported from outside are increased. The local villagers have to adapt to the dual markets, however, the tourist market, generated by modernization, is labeled as an official market and is gradually replacing the traditional local market.

*Commercial Sense*

It is quite striking how fast the local people adjust themselves to the social change. In the first investigation (2010), the local people showed less commercial sense compared to the second investigation (2011). During the first investigation, the author had to enter into local people's houses to do interviews, the local people, especially those who lived in the upper mountains were very friendly, and they always gave presents such as fruit and local products to us after the interviews. An obvious change during the second investigation was that the local people seldom gave us presents. Some of the elderly local villagers sold fruit near the tourist sites, the price was higher than in local markets and they even advertised that the fruits were healthy fruits without pesticide. This phenomenon never appeared during the first investigation. Just within one year, the local people quickly realized the value of their own culture and utilized the local products to help them earn more money.

*Increasing Income*

According to the third investigation (2015), when asked about which factors contributed to increased revenue, the result was more than 70 % of local people said the tourism economy. There are 53.6 % of local people said ICT helped them do business and 9.5 % said they used ICT to know the trend of the market. Improvement of tourism services and individual proprietorships are the main aspects that contribute to the fast-growing economy of the local villages. The three investigations all suggest that the local people who follow society's trend can benefit from modernization, since villagers involved in tourism have largely increased their revenue in the past few years. The tourism economy stimulates improvement of infrastructure and transportation, in turn, the better infrastructure and convenient transportation attract more tourists. The third investigation suggests that more than 80 % of people agreed that the local tourism economy benefited from the improvement of infrastructure and transportation. As the living standard of local people improves, the issue becomes how to distribute the gross income.

## Conflict of Interest Groups

The investigated region is mostly tourist regions which just opened their door to the outside recently. Within a short time, the gap between rich and poor and imbalanced development among different locations started to show. The local government of Xijiang decided to develop a tourism economy in 1995; at the beginning, most of the minority area lacked strategies about how to start. The local government of Xijiang and Zhaoxing made contracts with tourism companies, which meant the companies could make decisions about how to use the local culture. The problem was the companies were not local companies and the decision makers were outsiders with little knowledge of the indigenous culture.

The outsiders running the tourism companies brought devastating damage to the indigenous culture. For instance, in Zhaoxing, the old hospital was a traditional historical building, which was important to study the traditional architecture. However, the tourism companies tore it down and established a parking space for tourists. During the investigation, some elderly interviewees felt pity about it, and they regretted the loss of such a beautiful building, but felt helpless to protect their local buildings.

Places like Langde and Biasha are government tourism economies and compared to the private companies can better represent the local people's interest, however, keeping outside forces away makes it more difficult to develop the local tourism economy, since their financial problems restrict the development of a local economy. Currently, they are thinking about whether they should imitate Xijiang and introduce private companies (outsiders) to invest capital to develop the tourism economy.

Besides the strategies of developing a tourism economy, another issue is how to distribute the tourism economy revenue. The local people argue that they gave authority to private companies and governments to use the local culture and develop the economy, and indeed many tourists came to visit, but the reward was less than expected. Before the local governments and companies fenced off the village and set up ticket windows to charge entry fees, there was an agreement that the ticket revenue would be shared at the end of each year and each villager would get a share of the profits. However, except for the first year, when local people did receive a share, every following year the local people received little money or none at all. At the same time, the ticket office charged expensive entry fees which caused fewer tourists to enter the villages, hence, local shop and hotel owners got less revenue and the conflict between local people, local government and tourist companies became very acute.

When investigated deeper, it appeared the conflict not only affected tourist companies, local government and villagers, but also relationships among villagers, which became worse and worse. One typical case is the rapid increase in the divide between rich and poor among the local villagers. The inhabitants who live near the street can benefit more than the people who live in the upper mountains or who live far from the main street. It seems the tourism economy is an economy for a small group of people. The investigated region fully illustrates the routine of modernization and who can benefit from its fruit. During the modernization process, the homogeneity of society is increasing, for example, the phones we use are made by a few of the biggest smartphone companies and the cars we choose are also from the popular brands, therefore, the fruits of modernization are dropped into a few hands, such as giant International companies or countries.

## Social Mobilization

In the process of modernization, dramatic changes have happened in these places. In terms of social mobilization, there are two main compulsive forces that contribute to the social changes. One is oriented by the tourism economy, the other by ICT. There are different voices about whether new media tends to mobilize or immobilize their audiences. Some argue that the media, especially the modern media, bring malaise rather than mobilization to the public (Ansolabehere & Warburn, 1995; Blumler, Blumler, & Gurevitch, 1995; Kerbel, 1999). The sensational news from media or scandals from politicians might lead to "political apathy, alienation, distrust, cynicism, confusion, disillusionment and even fear" (Newton, 1999). The media, once better developed in big cities, has now spread

**Tab. 6-4:** Entertainment activities before and after tourism

| Before tourism | frequency/people | Percent | After tourism | frequency/people | Percent |
|---|---|---|---|---|---|
| chat | 420 | 44.7 | Chat | 329 | 35.1 |
| visit relatives | 350 | 37.2 | visit relatives | 256 | 27.3 |
| drinking | 247 | 26.2 | drinking | 209 | 22.2 |
| traditional recreation | 230 | 24.4 | traditional recreation | 183 | 19.5 |
| card games and mahjong[a] | 131 | 13.9 | card games and mahjong | 124 | 13.2 |
| local market | 308 | 32.7 | local market | 264 | 28.1 |
| Singing and dancing | 330 | 35.0 | Singing and dancing | 337 | 35.9 |
| watch TV | 321 | 34.1 | watch TV | 381 | 40.6 |
| reading | 130 | 13.8 | reading | 154 | 16.4 |
| learning handicraft | 176 | 18.7 | learning handicraft | 238 | 25.3 |
| sport | 167 | 17.8 | Sport | 191 | 20.4 |
| do nothing | 88 | 9.4 | do nothing | 90 | 9.6 |
| others | 34 | 3.6 | Others | 43 | 4.6 |

Source: Author's calculation.
[a] 麻将 also spelled Mahjong, very popular game in China.

very quickly throughout local villages and opened a window for local people to know what is happening outside.

The tourism economy mobilizes local people to show their traditional culture where they can earn money by dancing and playing local music for tourists. At the same time, modern entertainment such as karaoke, bars, or traditional performances occupies local people's activities. Putnam holds that media tends to privatize people and isolate the masses from the community and public associations. The consequence is it can cause "the decline of social capital", which is also attributed to "civic disengagement, loss of community and privatization of modern life" (Newton, 1999). Tab. 6-4 shows the entertainment activities of local villagers before and after tourism. It indicates that watching television became the first choice of entertainment. Before the tourism economy, local people liked to chat in the *fengyuqiao* (风雨桥), which is a traditional public space for local people to chat. Local people liked to meet outside and visit friends and relatives to communicate and have recreation together, which I designated outdoor entertainment. However, based on the past three investigations, more and more people prefer to stay home and watch television and use ICT to entertain themselves, which means the outdoor entertainment has been replaced by indoor entertainment.

Modern values and ICT change the minority villages dramatically. As the media, especially television, becomes the main part in people's daily life, for the local minority villages it is more "family entertainment" in the house than group entertainment outdoors. In the traditional Chinese society, relationships between individuals were considerably strong and lasting. As a local villager said, *"After the tourism economy, our village changed dramatically. In the past, many people go to the outside to work (dagong 打工), now they have returned. The economy has gotten better, in the past, we didn't have television, smartphone and computer, now we have. The bad thing is our river is polluted. Before it is very clean and you can see bottom of the river, but now it is very dirty. In the past, during the festival, we liked to see the cows fight. There were many people, all villagers came outside and watched the fight. It is very boring to see cattle fight, I like to see buffalo fight, buffalo is very aggressive. Now, there are not so many people have cows, because cows cannot earn money. Since late 1980s, the festival got less people to participate. Local people went to the outside to work.*[67] *Before our basketball playground was surrounded by people, after dinner, we rushed to the playground to occupy a place to see basketball. Now people who worked outside returned back, but they don't have an interest in our traditional entertainment now, they like to watch TV, play mahjong, play computer, and everybody holds their mobile phone. Young people just think about how to earn money fast and farmland is abandoned"*. (Song, 42, male, Xijiang, personal interview, July 30, 2010, 1B3)

As I have mentioned in the part on political change, the widespread of ICT has less influence on the political engagement of the masses, and the local government rarely communicates that way. One of the important reasons could be attributed to the villagers' low education background and strong dialect, the fact that only youth have a good command of ICT, and that villagers over 40 years old prefer to use old media. Since television is the main access for local villagers to receive information, in order to find usage differences by gender, age and education background, I designed questions such as: How many hours do you spend watching TV? What are your favorite programs? Etc.

The survey shows that women spend more time watching television programs such as drama and entertainment programs, and men spend fewer hours with

---

67  I have explained the urbanization process in Chapter 3.

**Tab. 6-5:** Most Favorite TV programs

| Most Favorite TV programs | | | | | | | | | |
|---|---|---|---|---|---|---|---|---|---|
| Gender | News | Television drama | Children programs | sport | entertainment | discussion | law | Science technology | service |
| Men | 36.3 | 12.8 | 2.5 | 4.3 | 2.5 | 1.0 | 2.5 | 1.5 | 1.3 |
| Women | 9.3 | 15.5 | 2.3 | 0.3 | 4.3 | 0.8 | 1.0 | 0.3 | 10.3 |
| <10 | 0 | 0.7 | 0.5 | 0 | 0 | 0 | 0 | 0 | 0 |
| 10-20 | 6.6 | 13.0 | 2.7 | 2.9 | 3.9 | 0.7 | 0.7 | 0 | 0.7 |
| 21-30 | 10.8 | 6.1 | 1.2 | 1.2 | 2.0 | 1.0 | 1.7 | 0.7 | 0.2 |
| 31-40 | 13.0 | 4.4 | 0.5 | 0.7 | 0.2 | 0.2 | 0.7 | 0.5 | 0.5 |
| 41-50 | 5.1 | 1.7 | 0.2 | 0 | 0.2 | 0 | 0 | 0 | 0.2 |
| 51-60 | 4.6 | 1.0 | 0 | 0 | 0 | 0 | 0 | 0 | 0.5 |
| 61-70 | 3.2 | 0.7 | 0 | 0 | 0.2 | 0 | 0.2 | 0 | 0 |
| 71-80 | 1.5 | 0.2 | 0 | 0 | 0 | 0 | 0 | 0 | 0 |
| >81 | 0.5 | 0 | 0 | 0 | 0 | 0 | 0 | 0 | 0 |
| Primary school | 43 | 39 | 7 | 0 | 4 | 1 | 2 | 1 | 1 |
| Junior high school | 42.6 | 38.6 | 6.9 | 0 | 4.0 | 1.0 | 2.0 | 1.0 | 1.0 |
| Senior high school | 45.2 | 29.6 | 6.0 | 6.5 | 6.0 | 1.0 | 1.5 | 1.5 | 1.0 |
| Vocational | 43.9 | 14.6 | 1.2 | 8.5 | 11.0 | 4.9 | 8.5 | 3.7 | 2.4 |
| Undergraduate | 57.1 | 14.3 | 0 | 0 | 9.5 | 4.8 | 9.5 | 4.8 | 0 |
| Graduate | 100 | 0 | 0 | 0 | 0 | 0 | 0 | 0 | 0 |

Source: Author's calculation.

television in general and prefer to watch TV in the evening, such as news, sports, law, discussion programs, and technological programs. Tab. 6-5 indicates men and middle-aged people are more interested in news and technological programs and that the better-educated villagers prefer to watch news and law programs.

Social scientists believe that media can stimulate the masses to participate in political and social issues. The media gives a lot of information to their audiences, and provide different opinions as well, which decreases the knowledge gap and mobilizes the masses politically (Gunter, 1987; Kleinnijenhuis, 1991). Regarding the minority region, indications are that social mobilization cannot be simply generalized as to whether the widespread of ICT increases the mobilization or not. According to the field investigation, political mobilization is influenced by individuals' education and age. The young, well-educated people are more engaged in the political process[68].

With the widespread of mass media, the local villagers are well-informed, with different access methods to media and are able to transmit their voices to the outside. Before tourism, the minority places were all hollow villages. Currently, a trend of returning to the hometown is starting among the local people. More than 55 % of people claimed that they knew migrant workers who returned to the village.

The social mobilization of the villagers depends on the individual's age, educational background and so on. The old generations have many obstacles between them and ICT empowerment. To utilize ICT is one of the important skills of the information age, and ICT is endowed with power, and people who can command the modern technology tend to show high status in the social stratification. As a result, the role of the old generation, which was regarded as the authority group in the traditional society, is gradually reduced.

## Psychological Adaptation

As I have mentioned, the minority area has its own indigenous culture and did not interact frequently with the outside before the tourism economy. The local people use their own dialects and keep their own customs. With the tourism economy and widespread of ICT, their life is reshaping: modern music replaces traditional songs, local festivals are replaced by festivals from majority regions, the elderly people lose their status of representing authority, and traditional customs are less important.

---

[68] The hollow village is used to describe villages that only elderly, women and children live inside, men and young people are working outside.

The Chinese society is a "face society" (*mianzi shehui*面子社会), people care about how others look at them and how other people judge them. The indigenous place is an acquaintance community; people are familiar with each other. If something happens, the news will spread quickly throughout the local community. As a result, they formed a self-censorship system, which censors individuals who live in the community and expects them to follow traditional rules.

The local villagers are now more open to outside society and show more tolerance to deviance. The reason could be attributed to the widespread of ICT, and modernity's preponderance over the tradition which caused the old generation to gradually lose their authority. For instance, before the tourism economy, it was impossible for local villagers to accept women who have children or lose their virginity before marriage. During the investigations, the old generation always said the society belongs to young people now, nobody listens to them. The diffusion of media, as well as the returning young people who worked outside, are bringing different values and opening local villagers' horizons.

## *The Changing of Clothes*

The investigated villages have their own traditional clothes. The old generations still wear the traditional clothes while the young generation only wear them during festivals and performances. The production of traditional clothing is very complicated, a lengthy and expensive process compared to normal clothes. In the Miao group, the preparation of women's traditional clothes begins when they are infants. The clothes will be finished before girls get married and they will wear them as a bride. In Biasha, Pianpo, and Zhaoxing, the clothes are still made by the villagers. The traditional clothes do not use any chemical dye material and everything is natural. The blue color is derived from herbs and it takes three weeks for the dip-dyeing process. The first investigation suggests most of the local people wear standard clothes (81.1 %), except for the people of Biasha, which still wear their traditional clothing for daily life; other minority places only wear their traditional clothes when they have performances or traditional festivals (see Tab. 6-6).

At the threshold of tourism, it was difficult for the old generation to accept modern clothing. When they saw tourists wearing miniskirts or clothes that were very short, they were afraid the young generations would learn from them. During the investigations, some teenagers claimed that their parents and grandparents used to point to tourists who wore short trousers, skirts or shirts and asked them what did they think about the clothes. The results were always

**Tab. 6-6:** The frequency of wearing traditional clothes

| Wear traditional clothes | frequency/people | Percent |
| --- | --- | --- |
| Daily | 183 | 19.9 |
| Show | 252 | 27.4 |
| Festival | 690 | 75 |
| Go to other city | 49 | 5.3 |
| Rent to tourist | 20 | 2.2 |
| Other | 77 | 8.4 |
| Total | 1271 | 138.2 |

Source: Author's calculation.

the same, that the youth were not allowed to wear clothes like the tourists and the parents regarded these kinds of clothes very ugly.

However, after several years of tourism development as well as the influence of media, the young generation was quickly assimilated into the modern fashion style. The youth style their hair and wear modern clothes so that there is no difference between minority youth and their east coast counterparts. During the investigations, the youth attitudes were very optimistic toward their future; thinking of themselves as young and representative of the modern culture. On the contrary, the old generations believed they were out of date and had lost their authority in the face of modernity. They felt that society belonged to the young generation, since they were dropping behind the technological development. This suggests that the technologies endow higher status to the individuals who command them, i.e., the young generation.

## *The Shift of Authority and Taboo*

The minority villages have their own religions. For instance, Biasha people have special beliefs about trees; Miao people have their own butterfly legend and regard the cow as a holy animal; Kam people believe in ghosts and that a room has to be spiritually cleansed after a visiting couple stayed overnight, etc. However, such beliefs and customs are gradually fading away in local people's daily life.

In Zhaoxing, if tourists stay together in a local hotel as a couple, the owner of the hotel sometimes requests a high price if they want to share the same room. The higher price is used to "clean" the house as they think the room is polluted. After the couple leave the owner will hire a shaman to "clean" the room since the room is "dirty" in local people's mind. Before the tourism economy, the local people had strict regulations about this custom. However, as more and more tourists visit Zhaoxing, they have had to adjust the old tradition to the new situation. Some private hotels do not even ask a shaman to clean the room anymore. I interviewed a private hotel owner Li, she said she cares less about whether a couple live in one room or not, just so they pay her extra money and she can ask a shaman to clean the room.

In previous times, when people broke the local law, the judges were always the elderly authority person. For instance, in the Kam group, most of the judges who could make a decision and decide how to punish criminals were people who were older than 50 years and educated men. They had their own court, which was at the *gulou* (鼓楼), a drum tower with entertainment and judgment functions. In the summertime, the *Gulou* is always crowded with local villagers, chatting and playing games inside. In the past, when local people broke the law, they were judged by *gulou*. On the top of the *gulou*, there is a wooden stick. Criminals arrested by the local villagers were held inside the *gulou* for the whole night. On the second day, whether the criminals should be punished or not was decided by the direction of the stick. The punishment was normally 50kg rice, 50kg flour, 50kg alcohol, 50kg meat, and 50kg fireworks. Afterwards, the authorities would distribute the goods to the local villagers. Now when the local people have conflicts, they prefer to ask government officers (47 %) or police (24 %) for help (see Tab. 6-7). As mentioned before, news and law programs are very popular among the villagers, which teach them how to use laws to protect themselves and also encourages them to go to government or police stations to resolve their problems. The elder authority people only play their roles in traditional festivals.

When modernization enters into an indigenous culture, what is the reaction of the local people? Tab. 6-8 shows that 46.8 % of the local people agree that they benefit from tourism and 16.5 % of the people earn money from it. However, another concern is that tourism brings a gap between rich and poor. During the field investigation, in these minority villages, people who live near the main streets have many more advantages compared to those who live in the suburbs or up in the mountains. As a result, the villagers who receive less benefit from the tourism economy are very unsatisfied with the local government.

It is interesting to observe the attitude of the local people toward outsiders. What are their opinions about people who come from outside the villages to work there? Do they want to attract more tourists to come to visit their places? Tab. 6-9 shows that most of the local people welcome outsiders to work and start their business in the villages, which also increases their confidence in their hometown.

**Tab. 6-7:** People to ask for help when conflict happens

|  | Frequency | Percent |
|---|---|---|
| Police | 276 | 24.0 |
| Officer | 540 | 47.0 |
| Patriarch | 61 | 5.3 |
| Rich and capable people | 14 | 1.2 |
| Teacher | 23 | 2.0 |
| Compounding | 155 | 13.5 |
| Others | 80 | 7.0 |
| Total | 1149 | 100.0 |

Source: Author's calculation.
Note: It is a multiple-choice question.

**Tab. 6-8:** The opinion of tourism (multiple choices).

| Opinion of tourism | Frequency/ people | Percent/% |
|---|---|---|
| We benefit from tourism economy | 702 | 46.8 |
| I earned money from tourism | 248 | 16.5 |
| The gap between rich-poor got bigger | 267 | 17.8 |
| Destroying the environment | 161 | 10.7 |
| Too many tourists which disturb our daily life | 90 | 6.0 |
| Others | 32 | 2.2 |
| Total | 1500 | 100 |

Source: Author's calculation.
Note: It is a multiple-choice question.

**Tab. 6-9:** Opinion of outsiders who work in the villages

| Opinion about outsiders | Frequency/people | Percent |
|---|---|---|
| Occupy the local market | 112 | 6.7 |
| Disturb the local rules | 95 | 5.7 |
| Like to communicate with the outsider | 267 | 16.0 |
| Our better economic performance attracts them to come to start a business | 313 | 18.8 |
| Help to boost local economics | 549 | 33.0 |
| They make our life more content | 238 | 14.3 |
| Other | 92 | 5.5 |
| Total | 1666 | 100.0 |

Source: Author's calculation.

The process of modernization has challenged the old values and has partly changed the social order. The local people have experienced a fast change from a relatively traditional society to a modern society. During the transition, firstly, the local people questioned traditional culture and chased for modern values. Secondly, after they reflected on their own culture they treasured it again. But during this process, the traditional values became less dominating to the society than before.

*Changes of Customs*

Singing and dancing were very important in these local villages. People who could not sing had difficulty finding their partners. Before tourism and the wide diffusion of modern media, young people used to sing to find their partner. Men and women stood in a parallel line, first, a man would begin to sing, if a woman was interested in this man, she would reply to him with local songs. Currently, this tradition is vanishing. Young people use digital devices to contact each other, even the old generations no longer sing, they use MP3 or their mobile phone to listen to music when they work in the field. During the first investigation (2010) in Zhaoxing, a man who is more than 70 years old said "I always listen to music when I am walking and relaxing. Some of my friends don't know how to use a smartphone to listen to music. They all admire me because I can use it to play music. But I think that our young children being unable to sing our traditional songs is really bad. Our songs are on the list of world cultural heritage, after our generation dies, I think no one can sing our songs."

Since the old generations also spend a lot of time watching television, the modern media is also shaping their values. They spend a large amount of time with the media, as a result, they will question the values they hold and reflect on them. The process of modernization brings new values, a different sense of law, modern technologies and a new societal system, but also makes the local people more material-oriented with less trust about the local community.

## Live or Achieve—Who is the Owner of Modern Society?

In the investigated region, even though it is a minority village that has cultural attractions for tourists, the minority culture alone is not enough to attract tourists to create a year-round economy. The local people mobilized the wisdom of the masses to consider how to develop the tourism economy. They started to use air conditioning, even though the local area is very cool in summertime; they bought mahjong automatic tables, albeit mahjong does not belong to the local culture; they organized singing and dancing programs, even though the actors and actress are outsiders; and the karaoke, bars, jewelry stores, "local" food, etc. are not all from the local culture, but if the tourists like it, they will do it.

As one local villager said, "*Our village is deserted and the prosperity is just an illusion. Many outsiders come here to run businesses, but they don't belong to us. When the tourism economy is not promising, they will leave; the tourists just come here for a short time, they'll probably never come back again; the most important thing is the young generation don't want to stay. People who are forty or fifty maybe will still stay here, but if they have money, they will send their children to study outside, because to stay here provides no future. In the job market, the children have to compete with their majority counterparts who can speak Mandarin since birth, play piano and violin when they're young. What about our own dialects, traditional instruments, and songs, we can't use them to earn money.*" (Hou, 52, male, Xijang, personal interview, August 5, 2015, 14)

Such thinking is very common in these villages, which inspired me to think deeply about who is the owner of the modernization, and whether tradition will fully fade away under modernity. Based on the interviews, local villagers only regard the indigenous culture as a way to earn money. They have lost the inside connection with their traditional culture. As more and more outsiders come to their villages to work and occupy many of the job opportunities, the local villagers who lack commercial sense have to go outside to work. Therefore, the local area is like a banana whose outside and inside are different colors: from the outside, it still keeps its original architecture, but when the exterior is peeled away, the local people who are the carriers of local culture are gone. People who are left in the villages have to walk narrow mountain roads and use horses to bring heavy things because the tourism economy needs the local area to be maintained as a museum. However, during the modernization process, should the tourism economy deprive them of their rights to live in a modern house, walk on a flat wide road, and use vehicles to carry heavy loads? Is modern life an ideal life for the local villagers? During the modernization process, who plays the most important role to introduce modern values and how do the modern values spread to the local villagers? Before replying to these questions, I will start with a dialog with a local intellectual.

## A Dialog with a Local Intellectual

The local government is the main propellant for the tourism economy, people who work in the local government are intellectuals and well educated. The local intellectuals are very active and creative. While conducting field research in

Pianpo, a well-educated elderly man called Tingchao Chen[69] left a deep impression on me. He was 75 years old at the time and his family is one of the oldest local families in Pianpo. He is Bouyei minority and the former headmaster of a minority school. He had established a private library, which encouraged more local villagers to learn the traditional culture and he was working as a volunteer teacher of the local culture as well. Regarding the developing tourism economy, he believed the village needed new, bright shots to attract more people to come to visit. Compared to other tourist minority villages, Pianpo lacked enough indigenous culture and the landscape of Pianpo is also average. According to Chen, the new events should aim to introduce other cultures, which could be combined with their own culture. After consideration, he introduced and designed a Pangu[70] (盘古) ceremony in Pianpo.

There are reasons why he chose the Pangu ceremony. The legend of Pangu is part of the indigenous culture of Pianpo. Pianpo had an informal ceremony in the past, but it was loosely organized and uninformed. He referenced the Pangu ceremonies of Henan and Gansu provinces, and selectively chose cultural elements, which fit Pianpo and applied them to their Pangu ceremony. Even though he could not use the Internet, he could easily get information from it As mentioned earlier, the elderly people always live with their youngest son in the local village, therefore, his son could help him check information online. By collecting and selecting useful information, he combined the particular cultural elements with the local culture and created a new Pangu ceremony. Currently, the Pangu ceremony is an important part of the local culture in Pianpo and it is also a main event in the six-six ceremony (liuyueliu六月六节). When interviewed, he said:

> "The original idea of a Pangu sacrificial ceremony was my proposal. Travel to a place is not only about the landscape, but also culture. The tourism economy here is a bit boring, except for singing, we don't have other programs. I thought it does not work if we keep developing like this. We should do some changes and a ceremony would be a good idea. I discussed my idea with people who were a similar age to me, and they all agreed with the proposal. In my hometown, we used to have a Pangu sacrificial ceremony; however, it was just a superficial program under the name of Pangu, which lacked essential meaning. Bouyei people (the local villagers) like to plant rice, whereas Pangu is the ancestor of dragon's root (龙根)

---

69  Investigation number:2.1B1.
70  Pangu is a legendary person in Chinese mythology. He is regarded as the first human being who began creating the world. He separated the earth and the sky with his axe, which in Chinese means Yin and Yang respectively. According to the legendary stories, mountains, rivers, land, stars, forests minerals, rain, animals, etc. are all parts of Pangu. When he died, his body changed into nature.

*of rice. After the decision, I proposed we should do some research, collect materials, and design the procedure for the ceremony.*

*I am the organizer of the whole ceremony. I wrote the ceremony speech,* **my son helped me download the pictures online***, and I also organized villagers to participate in the ceremony. The procedure is as follows: (1), We sacrifice a pig and a rooster to Pangu.*[71] *(2), At the beginning of the ceremony, everyone should stand up as well as audiences.*[72] *(3), We beat the drum six times, play our local instrument (changxiao, Chinese: 长萧) six times, and shoot with a special gun six times. Six means everything will go smoothly and it is also the date when Pangu died, because of this reason, the festival is also called the six-six festival (liuyueliu六月六). (4), When we finish all of these procedures we say together: liuliu dashun (六六大顺，means everything goes well). We pray our plants will have a good harvest, the weather will be good, and all our villagers will be healthy. (5), There are 10 young boys and 10 young girls who read poems together to carol Pangu. (6), There are 24 adults wearing yellow scarves on the stage to burn incense, bring holy water and holy soil to Pangu. The holy water is chosen by the wells which are located in the south, north, west, and east; the four directions of the village. We call these wells dragon wells. The water is drawn in the morning to guarantee the water has not been used in the new day. We believe it is important to use pure water to celebrate the festival. The soil is picked from caves where nobody has worked on the soil before. Similar to the holy water, we picked the holy soil from the four directions as well.* **The procedure of using the holy soil is learned from Henan province. Henan province has celebrated the ceremony for more than eight years. I learned it from the Internet. The sacrificial ceremony was learned from Gansu province through the Internet, and they use this way to perform the Yellow Emperor (huangdi**黄帝**) ceremony.**" (Chen, 75, male, Pianpo, personal interview, August 9, 2011, 2.1B1)

The above description about how the Pangu ceremony was established implies the relationship among the intellectual, media and social change. Chen, as an intellectual who is deeply familiar with the local traditional culture, knew what the local village needed and he knew which new culture would be a better fit the indigenous culture. When standing on the wave of modernization, intellectuals such as Chen reflect on and are aware of the indigenous culture. As he said:

*"I am an intellectual. My interest is in how to develop our local culture. I learned a lot about the traditional culture and I believe I am a qualified teacher. The Bouyei language is like the Latin language, which is very difficult to learn. I have encouraged the local students to learn the Bouyei language, however, most of the people think it is useless to learn and don't have motivation. I regret very much that I did not insist on teaching the*

---

71 The meat which was used for sacrifice is very popular after the ceremony. Everybody wants to eat it, as they believe that eating the meat can bring them luck, happiness, health, and their children will be smart and the elderly will have a long life span.
72 There were more than ten thousand people in the audience participating in 2010.

> local language, since we will probably lose our own language now. At the moment, the tourism economy gives me hope, it seems it motivates more and more youths to learn about the traditional culture. In the local school, there are teachers teaching the local language, local culture and many teachers teach for free." (Chen, 75, male, Pianpo, personal interview, August 9, 2011, 2.1B1)

The local intellectuals not only have deep recognition of traditional culture, but also play a role as organizers. They are willing to participate in politics and would like to apply their knowledge to the society. By mobilizing the masses, they gladly dedicate their knowledge to the society. "*In 2008, I (Chen) established an elderly people group. Every year I organize the group to go outside for travel. Currently, we have 30 memberships. The aim of the group is to go outside and communicate with the outside; experience other places; support our government's work; spread our traditional culture and help each other. One time, local people had a conflict with the government because of farm field issues, so we coordinated for both sides and solved the conflict. There was one family whose house was destroyed by a fire, we encouraged villagers to donate money and material to the family and helped them pass the difficult time. Besides these kinds of things, we also helped the young, single men find girlfriends. We have meetings regularly and we also give feedback and suggestions to the local government.*" (Chen, 75, male, Pianpo, personal interview, August 9, 2011, 2.1B1)

When asked about his family inn, he said "*We borrowed 3500 dollars from a bank to invest in our family inn. We bought a computer and built a bar, we also want to buy an air conditioner, however, the wooden house restricts the use of other electrical appliances, so it is unrealistic to use an air conditioner. Meanwhile, we bought mahjong and karaoke machines, these machines largely improved our revenue; the tourists that come here also want some entertainment.*" (Chen, 75, male, Pianpo, personal interview, August 9, 2011, 2.1B1)

Similar to Chen, there is another person playing an important role during the process of modernization. He is called Yang and is the leader of Pianpo village. As a Han (汉) person[73] working as village leader, he began to realize the indigenous culture is special. As an outsider, in order to develop the tourism economy and improve the cultural identity of the local village, he initially wore the minority clothes and encouraged local villagers to wear the traditional clothes as well. He learned the dialect and sang the local traditional songs. In addition, he used media to make an advertisement for the Bouyei (百越) culture, at the same

---

73 Han people are the majority ethnic group, representing about 92 % of the people in China.

time he encouraged the local people to protect the culture and realize their culture is a treasure. He set up regulations about how to protect the local architecture: for the buildings which still exist, the local people should keep the original appearance; for buildings established in the future, they should strictly follow the building regulations.[74] In order to encourage the local people to develop tourism, he suggested the government give a prize to the first eight people who start a tourist business. As an outsider, he could observe and appreciate the character and values of the indigenous culture.

The above description of the role of local intellectuals highlights their utilization of the media to collect information, combine it with their values, and create something special for the local people. On the one side, they intensified the indigenous culture, on the other, they changed the local culture to make it better fit the modern world.

Why are the headmaster Chen and the local leader Yang, as intellectuals, focused on spreading their traditional culture and very open-mindedly selecting some values from modern society to develop the indigenous society? Why are they keen to mobilize the masses and play a role as an organizer? What is the role of intellectuals in China?

Similar to the intellectuals in local villages who introduced new culture to their indigenous culture, in the beginning of the 20th century, some intellectuals who studied or worked in Western countries believed that Western values and modernization could lead China to independence and prosperity. To achieve this target, the media were the main tools for them to spread their values and their main access to new values from foreign countries.

In the investigated region, intellectuals apply and spread their ideas to the local government and villagers with the help of media. They are the main connectors of knowledge to the masses. When the Internet started to be adopted in China, scientists and people who were well educated were the frontier users. The intellectuals as the brains of the society and the main force that use the communication system to spread new values and concepts to society.

## Conclusion

In this chapter, I introduced the investigated area and used four aspects; economy, social structure, politics and values, to explain why I regard the minority region as a traditional society. This area is experiencing a fast social change which is

---

74 The regulations are published in the center of the village so that everybody can check them.

driven by the diffusion of ICT and a tourism economy. The transformation of the villages explains how the indigenous culture is changed by modernity and in turn reshapes local people's self-reality.

Before the tourism economy, the local villages changed slowly and mainly followed their own customs. However, since many tourists have rushed into the villages, the local people who worked in the big cities have returned to their hometown to do business, and at the same time infrastructure was built and the pervasive diffusion of ICT has dramatically changed these indigenous villages, including their languages, clothes, buildings, festivals, customs, entertainment, and so on.

The diffusion of ICT improves the knowledge of local villagers. As the young generation widely use the new technologies, they learn Mandarin by using smartphones and computers. The rushing tourists and media bring different lifestyles and values to the local village, which challenge their indigenous culture. The change of communication system also shapes the relationship between the local government and the masses. The local villages have to live with dual markets, at the same time, their commercial sense has dramatically increased as well as their income. The economic change is accompanied by a rich and poor gap and a geographical divide. For the local people who are involved in the modernization process, they have to psychologically adjust themselves to the social change.

Tourism stimulates the process of modernization and accelerates the diffusion of modern technologies, especially information communication technologies. The ICT plays a crucial role in the tourism economy as well as in communication between the government and the masses. The government uses media to spread the information that the villages are attractive and worthwhile to visit; at the same time, the government also pervades and teaches the masses how to utilize their culture to earn money. They propagate their values, and consequently, these values internalize into the local people's self-reality and guide their activities.

The intellectuals play a role in introducing the new culture, reflecting on the traditional culture, bridging traditional culture and modern values, and thus selectively choose new cultural elements to enrich the traditional culture. In terms of the investigated villages, Chen and Yang are both intellectuals of the village. When modernization began, they reflected on their traditions and used media to selectively choose parts of exogenous cultures to combine with the indigenous culture. The local intellectuals selected cultural components that were similar to their indigenous culture and applied those that fit the local area, selectively absorbing elements of the chosen culture. As a result, a new culture is generated and spread to the whole society.

# 7 Reflections on Indigenous Culture and Modernization

At the threshold of the Chinese Economic Reform, Deng Xiaoping proposed to let a small group of people get rich first, and then have these people help the majority get rich later. However, the current situation seems to have not turned out as he expected, since a small group of people immigrated to countries with better living standards and transferred their wealth outside. At the same time, the fast modernization process caused the society to become money-oriented: adults go to urban centers to work resulting in millions of children being left at home to be taken care of by grandparents or some youth homes. For the children, they communicate with their parents by mobile phone, and their friend is ICT.

Modernization as an indigenous culture in the Western countries developed gradually, whereas for the laggard countries such as China, the wish to use modernization to improve the living standard quickly invoked societal changes very violently. Can modernization be a blending process that welcomes every traditional culture to join? If so, will modern values become dominant so that countries involved in the modernization will experience a homogenizing and irreversible process? If human beings are carriers of culture and connectors between the values of modernity and tradition, can we regard the elderly people as carriers who represent traditional culture better than the young generation?

As mentioned earlier, modernization in the West progressed concurrently with the indigenous Western culture. Western societies have witnessed many centuries as the modernization process developed into a relatively stable status. In the meantime, the educational level of individuals increased gradually along with their living standard. Compared to the large usage gap regarding modern technology adoption in the laggard modern countries, the gap in the West is small.

The laggard modern countries regard the Western countries as examples. However, the catch-up countries, especially their social systems, are lagging behind the technology development and the educational level of individuals remains relatively low. The diffusion of ICT is an enormous force that changes the four dimensions of the society, namely political, economic, cultural and structural. During the modernization process, these four aspects developed unequally, and in the case of China, the economic development is ahead of the other three aspects. Therefore, what are the consequences of the fast modernization process?

## Behind the Fast Speed of Economic Growth

In the context of modernization theory, the developed countries have been considered as precursors to spread new technologies. Countries like England, the United States, and France are regarded as the societies of the indigenous developers of modernization. "Other societies are the societies of late-comers. German society was, of course, a relatively early late-comer society; the Russians were somewhat later late-comers; the Japanese still later" (Levy, 1966, p. 12). The good economic performance of Western countries reinforces the belief that the liberal democratic system and individualism are good for the society. For other countries, the prosperity of the West leaves an impression that individualism, along with a liberal democratic system, is the way to a better life. For the individuals who live in this system, as most people follow the system, it is easier to live inside rather than doubt its legitimacy (Albarran, Chan-Olmsted, & Wirth, 2006; Humphreys, 1996; Schramm, 1964).

Nature likes diversity, not modernization. "Universal melting of identities, dispersal of authorities, and growing fragmentariness of life which characterizes the world in which we live" (Bauman, 1997, p. 211). The essential feature of modernization is similarity. People like to buy cars, clothes, digital devices, food and so on from several well-known brands, therefore, some companies grow fast and become big monopolies. People prefer purchasing the same products, and indeed, this makes the society more efficient—since the beginning of modernization, human beings have achieved much more than in the past. Besides the products, Western values established on the strong economy have persuaded the laggards to believe that modernization can help them rise above poverty.

For the minority places, in the past, the local people kept their indigenous culture and had less interaction with the outside. However, nowadays, they want to live modern lives like what the media shows them; they want to participate in the modernization process; they want to have a better living standard, and they want to follow fashion and through education become modern men. According to the third investigation, 14.3 % of people chose that the village before tourism is much better than now; 48.8 % local villagers said they prefer the current village, and 35.7 % said urban cities are better. Based on the attitude of the local villagers, it suggests that most of the villagers like the modern life. However, is the indigenous culture prepared to transform into a modern society that originally came from the West?

In Xijiang, the only high school is under pressure to move to another place that is far away from the village, because a tourist company needs the ground. Besides, the students in Xijiang high school are exposed to noise as the performance stage is directly opposite the school. Teachers are busy earning money and less focused on their job, therefore the commercial atmosphere makes the youth less focused on knowledge, but they can start to earn money at a very young age. The quotes below are from the investigation of the villagers:

> "Students here are bad quality. Good teachers switch their job to another town or a better place. Look at my daughter, as a high school student, she doesn't have any prize or award. If you, as a mother, do you buy a mobile phone for her? If I buy a mobile phone for her, she probably just gets 10.[75]" (Yang, male, 43, has a 15-year-old daughter and 12-year-old son, neither of them have a smartphone, personal interview, August 16, 2010, 1.2B2)
> "The children here always drop out of school very early (junior high school), they are influenced by the tourism economy. Students who have mobile phones always have bad performance in schools. When they graduate, they won't choose further education like go to technical schools. Boys will go to work (dagong 打工), girls will find men to marry. The mobile phone is indeed convenient, but our place cannot compare to big cities. Children in big cities, always have better education, their parents have more time to take care of them. But we don't know how to educate our children and don't have enough time. We didn't have a good education before." (Dong, 28, female, Xijiang, personal interview, August 14, 2010, 1.2B9)

The development of the indigenous society is unbalanced. Except for the pervasive ICT, the indigenous society lacks a proper education system. The educational background of the local villagers is low; the political system is unstable and the masses largely depend on the local government to make decisions whereas the local government has fraud, corruption and counts on tourist companies to pay taxes to balance the administration cost. Moreover, the places also face a laggard social welfare system and brain drain of the local villagers. The destiny of the local area is not controlled by the local people, but controlled by the local government, the tourist companies and most importantly, the people who follow the trend to harness modern technology.

## The Digital Divide: The Usage Gap and the Age Divide

During the third investigation, more than half of the villagers held that their lives are influenced by the tourism economy, followed by ICT. The pervasive diffusion of ICT decreases the physical gap, however, can the local people 'use' the technology? As the world enters into the information age, internet skills are considered vital assets. After the widespread of physical access, more studies suggest that the skill divide has become the more acute issue rather than the physical access

---

75  Chinese test full mark is 100, the pass score is 60.

divide. In developed countries, physical access gaps are decreasing, whereas the usage gap tends to grow, intensifying the existing societal inequalities (DiMaggio & Hargittai, 2001; Schradie, 2011; J. A. Van Dijk, 2005).

What is the usage gap? According to Van Dijk (2011), the usage skills include four aspects, namely mental access, material access, skills access and usage access. Afterward, Van Deursen and Van Dijk (2011) elaborated on the range of internet skills by proposing:

- *Operational internet skills*. These are derived from concepts that indicate a set of basic skills in using internet technology.
- *Formal internet skills*. These skills relate to the hypermedia structure of the internet, which requires skills of navigation and orientation.
- *Information internet skills*. These skills are derived from studies that adopt a staged approach in explaining the actions via which users try to fulfill their information needs.
- *Strategic internet skills*. Users with these skills have the capacity to command the internet as a means of reaching particular goals and for the general goal of improving one's position in society. The emphasis lies on the procedure through which decision-makers can reach an optimal solution as efficiently as possible.

Adapted from Alexander Van Deursen and Van Dijk (2011).

During the third investigations, the questionnaire looked at the relation between education level and the ability to use ICT to get information: the results suggest that the young local villagers with better education tend to use ICT to get information, 53.6 % of local people said they benefit from ICT to earn money, and 9.5 % people use ICT to check the trend of the market. The skill of the Internet is in the operational dimension, and the local people lack trust in E-commerce, since 70.2 percent of the local villagers don't trust online shopping and 82.1 % of them prefer to shop in real stores.

As the Internet starts to diffuse, all the young students have to learn computer skills as written in the education policy, which is as important to learn as the mother tongue. The new generations can grasp information and learn new technology faster than the old generations. They are cultivated with the capability to adapt to the information explosion age and distinguish various forms of information. Based on previous studies about the digital divide in Western counties, the age usage gap will decrease with time. But in China, the age gap is the most crucial usage gap and the gap will become bigger and bigger and will continue to grow. The information society has separated individuals into two parts: information inclusion and information exclusion, the elderly generation are excluded by ICT.

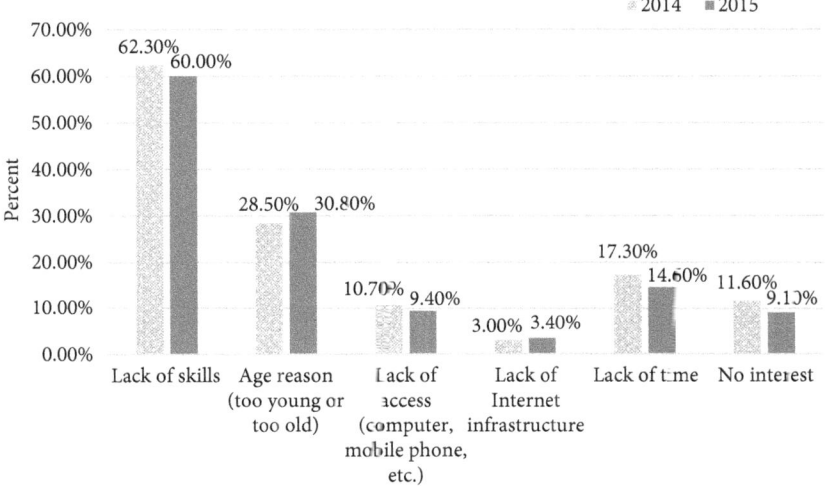

**Fig. 7-1:** Reasons to not use the Internet
Source: CNNIC (2016).

According to the 37th report from CNNIC, it shows that the growth of new Internet users has slowed. In the foreseeable future, the potential new Internet users are getting less. The report from CNNIC indicates that among the Non-Internet users, there are 11.8 % of people who indicated that they might use the Internet in the future, however, there are 72.9 % people saying that they will not use the Internet. In terms of reasons why people do not use the Internet, as can be seen in Fig. 7-1, compared to 2014 age is the only factor still growing.

Fig. 7-2 shows that people between 10 to 39 are the main Internet users. People who are older than 50 show less will to use the Internet, moreover, among them those who can use the Internet are just minorities which account for less than 10 % of the whole of Internet users.

The Cyber–optimists believe the penetration of ICT might melt the crystallized Chinese society and shrink the gap between rich and poor as well as different regions, but the age digital divide will become a big problem in the future. Given that the age gap seems to be the main digital divide in China, the consequences it brings is different from the digital divide in the West.

As ICT continues to spread, it changes traditional ways to record social memory and leads society into a new era. The country known for its long history and deep Confucianism influences may not exist as the diffusion of ICT and the

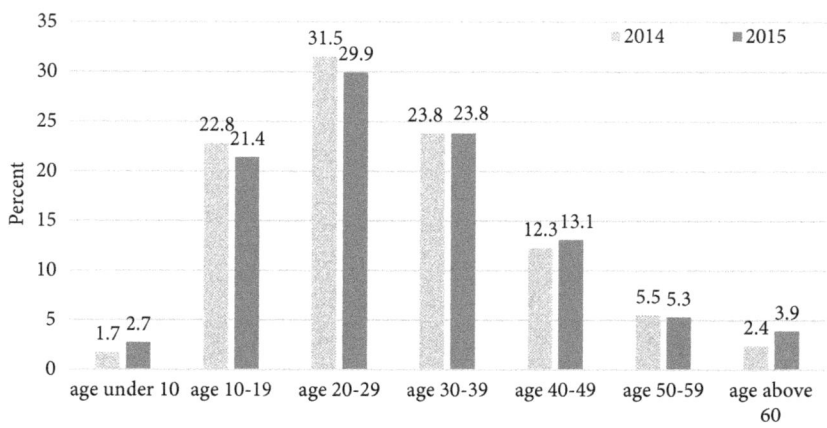

**Fig. 7-2:** Age of Internet users
Source: CNNIC (2016).

age divide continues. China is experiencing fast modernization, though the old generations, who represent the past social memory, are still the main carriers of traditional culture. Since the old generations participate less in ICT and are forced to become the silent group,[76] the current social memory is dominated by the young generation, leaving a disconnect between past and present social memory.

## Different Carrier— Tradition and Modernity

As mentioned above, one of the consequences of ICT diffusion and fast modernization is an age divide that results in the present social memory being loosely connected with the past social memory. If the old generation represents tradition and the young generation represents modernity, since people are carriers of the culture, how do the two carriers develop during the interaction process of tradition and modernity?

Given this question, I added questions in the third investigation questionnaire, such as do you think you are a traditional person? Do you think people who work outside are more modern than you are? For youth, the question was designed as, "Do you think your parents and elderly people in your village are

---

76  One of the reasons because of Pinyin which I have mentioned in Chapter 3.

traditional persons?" For adults, "Do you think your children are more modern than you?" The results suggest that local people who are over 35 years old tend to identify as traditional. There are 50 % of the people chose that people who work outside are more modern than people who stay in the village; 94 % thought that parents and elderly people belong to the traditional group, and 69 % chose children as being more modern than their parents.

The first two investigations suggested that the old generation rarely use new media. To more deeply understand the influence of ICT on the traditional culture, in the third investigation I tried to find out how do the young generations get to know the traditional culture. It shows that 45.2 % of local villagers indicated they get traditional knowledge from the older generation, 17.9 % from their family members, 17.9 % chose local government, 3.6 % and 1.2 % said local schools and the Internet respectively, and 14.2 % said from other sources. This result shows that the old generation and family are two important means for the young generation to learn about the traditional culture.

When asked whether media helped to protect the local culture, 60.7 % of villagers responded positively, but the local villagers have complicated feelings about ICT. On one hand, ICT stimulates the young generation to learn Mandarin, whereas local dialects are only spoken by the old generations. As more and more of the young generation work outside, the local dialects will disappear soon. On the other hand, ICT is the main platform to record the indigenous culture and are widely used by the young generation. The usage of ICT decreases the distance between young generations in the local area and young generations in urban areas, but it increases the distance between the old generations and young generations in the local area.

Based on the first two investigations, I assumed that the young generation and the old generation use different media to get information. In the third investigation, I designed a question asking, "Where do you get information?" The answers suggest that 88.1 % of people think the old generations get information from television and the young generations get information from the Internet. According to the Dependency Model of Mass Media, people who have different informational access tend to have different values and social realities.

The old generation is the carrier of the traditional culture and the young generations largely depend on the old generation to learn the traditional culture, however, the two culture carriers have different accesses to information. The problem is the modern media as the main channel to record social memory is now dominated by the young generation, whereas the voice of the old generation is forced to be silent. It is generally believed that physical access in the developing countries is still widening and is the main cause of the digital divide (Brandtzæg et al., 2011; Robison & Crenshaw, 2002; A. van Deursen & van Dijk, 2010).

In the developed countries, the information-not are people who do not want to use ICT, not because they cannot get access. However, the information-not in China are various. They may lack good education or physical access, have a language barrier, or have a typing problem, etc.—all these reasons can make it difficult for them to command ICT. In China, the age digital divide is the most serious consequence of modernization, resulting in the current social memory partially losing its connection with the past social memory. Therefore, accompanying the fast modernization process is part of social memory vanishing, as the old generation cannot participate in the modern technologies; they lack access to interact with the young generations, and most importantly, they will not wait there forever.

## Cultural Awareness

As Chinese society is faced with gaps of rich and poor, urban and rural, and east and west, at the same time, the age digital divide has loosened the connection between the current and past social memories. In the wave of fast modernization, will China follow the same road as previously modernized countries, and will the homogeneity of modernization assimilate the traditional Chinese culture?

Studies have proven that a strong economy is an important factor to intensify an individual's confidence in their own social system and ideologies they have adopted. A good living standard, stable life, and good welfare systems have had fundamental influences on the cultural identities of individuals(O'Shaughnessy & Stadler, 2012). After the investigated areas were opened up to the tourism economy, the many tourists coming to the villages made the inhabitants start to reflect and become much more confident about their local culture (see Tab. 7-1).

With the modern ICT, the local villagers can compare their own culture to other cultures and have new cognition toward their own culture. Tab. 7-1 above suggests that 51 % of the people feel happy to show their traditional culture to tourists and 39 % of the people feel very proud to show their culture.

Fei Xiaotong (费孝通), as a pioneering Chinese researcher and professor of sociology and anthropology, originally introduced the concept of cultural awareness (文化自觉), wenhua文化 means culture, zi 自 means oneself, jue 觉 means realize. In the 1980s and 1990s, he investigated many places in the minority areas of China and collected large amounts of materials. He pointed out that it is very important for minorities to reflect on their own culture, especially in the process of modernization. When the society begins to transform into an information society, it will accelerate the cultural change, hence, one of the crucial questions in front of countries that are involved in

Tab. 7-1: How local villagers feel about showing their traditional activities to tourists.

| Feeling | Frequency | Percent |
| --- | --- | --- |
| Just performance, only for earning money | 60 | 5.9 |
| Very proud to show their culture | 395 | 39.0 |
| Very happy and enjoyable | 518 | 51. |
| No special feelings | 119 | 11.7 |
| Total | 1092 | 107.7 |

Source: Author's calculation.

the modernization process is how to protect their traditional local culture. Fei claimed that when traditional culture cannot provide a comfortable lifestyle and cannot guide the activities of individuals, and meanwhile, a new culture is generating, during this process, reflexivity is the first try of cultural awareness. In the process of cultural transition, individuals are all involved in the transition procedure (Fei, 1997).

Cultural awareness means people (insiders) who live inside of a certain culture can reflect on and deeply consider their local culture. The insiders should face and reply to such questions as: Where does the traditional culture come from? How is it formatted? What are its characteristics? What is the future of the traditional culture? and so on. The reflexivity does not mean cultural recurrence or cultural instauration; it does not mean to abandon traditional culture in order to accept new culture or entirely accept Western culture. Insiders are actors who live in the environment of traditional culture, which is required to have self-awareness capability during the transition process. Self-awareness is the first step to adapt to the transition of society, and in the process of modernization it decides how the traditional society interacts with the modern society (Fei, 1997).

Another concept that is closely related to cultural awareness is self-awareness. It means individuals as autonomous beings can make their own decisions when outside culture enters the indigenous culture (Fei, 1997). Self-awareness is a very complicated process. First, individuals should deeply understand their own culture, only in this way can they find the position of their own culture among other multi-cultures. Secondly, when intertwined with other cultures, they can complement one another and reach the final aim of cultural awareness, which is to establish an order in which different cultures can get along with and respect each other. During the modernization process, individuals who are involved in modernity will gradually become aware of their indigenous culture.

The below interview indicates the cultural awareness of the indigenous place. When I interviewed Li,[77] who owns a handmade product shop, she said "*You run around in this street many days, I guess you probably cannot collect useful information since in this street there are many people selling clothes and local products but they are not local people. These are people who don't know our situation. The shops which sell silver products are mostly local people as well as those who don't have shops, just sell products near the street, they are local villagers.*" (Li, 40, female, personal interview, August 6, 2011, 2.3.12)

When talking about the division of traditional Miao culture, she mentioned Professor Louisa Schein,[78] she said "*I know her and saw her, she is a French. She indeed dedicated a lot of time for our local culture, however, she brought our best embroidery to France, now if you want to buy the best embroidery products you have to go to France. It was in 1995 or 1996, she suggested Xijiang should develop a tourism economy, afterwards, she contacted collectors from France who came here to buy embroideries. At that time, one piece of embroidery was 30 dollars. Do you know how much it cost now? At least 120,000 dollars! We cannot say it was cheating, because at that time these old embroideries were inherited from old generations, most of the people thought it was useless. They stored the embroideries at home and nobody wanted to wear them. These embroideries were old and we all sold to her. We thought these embroideries were useless but for them these products were like big treasures, because we were not aware they were so valuable. At that time, I did business about collecting embroideries, we had competition for collecting embroideries with the French collectors. When I said 30 dollars, they would say 45 dollars, therefore, it was very hard for me to bid with them. I was very shocked at that time how the embroideries could be so expensive, but I could not imagine the current price[79]!*" (Li, 40, female, personal interview, August 6, 2011, 2.3.12)

When I asked why the embroideries are so special, she said "*Good embroideries, the fabric and color are all self-made. We used banlangen* 板蓝根（*roots of*

---

77 She is 40 years old. She has experienced many different businesses. Now, she owns an embroidery shop. Interview number:.2.3.12.
78 She is an anthropologist and currently working in the department of women's and gender studies in Rutgers University in the United States. She can speak the local dialect and has done much research in this region. She is well-known among the local villagers in Xijiang.
79 To confirm the information, I contacted Prof. Schein, she replied she did not introduce Miao embroidery to France. It may have been Gail Rossi, a collector. But this sounds like legend either way.

*Isatis tinctoria*) *and other herbs together to make color. Now it is difficult to find the herbs, now there are mass produced products and the values of embroideries is of course lower than before. When the collector from France came here, they collected embroideries from past generations, these embroideries are at least two to three hundred years old. The collected embroideries required the embroideries cannot be washed. Our Miao people wear embroideries in marriage and death ceremonies, during the ceremony or festival, we wear the best embroideries to show our respect."* (Li, 40, female, personal interview, August 6, 2011, 2.3.12)

When talking about passing on the culture to the next generation, she said *"Now the young generation don't know the traditional skills, they all work outside (dagong* 打工*). People who are left are just elderly and children. It is difficult to teach the young generation the traditional culture. The most important thing is the influence of the one-child policy and modern culture. Think about when you have a daughter, you can teach her embroidery skill. But in our region, we think boys are more important than girls.*[80] *The Internet, television and so on distract them, and gradually this skill is in danger. If I have enough money, I will definitely buy all embroideries and bring them back. I admit that compared to the local people, the foreigners were aware of the values of our local culture earlier than us. I feel sad because the embroideries will never come back. What they cannot say is cultural invasion, but they utilized our ignorance, their aim is reasonable, but the way they used is not good."* (Li, 40, female, personal interview, August 6, 2011, 2.3.12)

When asked her opinion about the local tourism economy, she said *"People all said our place has too much commercial atmosphere; I think because of the local government. They are not dedicated to their job, just boast to themselves about the local economy. They lack the capability to plan a tourism economy. I think we should keep our indigenous culture, because the commercial stuff you can find everywhere, however, the Miao minority in Xijiang is special. Now when you pass the ticket station, the first thing you see is the people who work as security guards, city officers and police. Is this Xijiang? So many obstacles give an impression that people who live inside are criminals. Why do these people wear uniforms? These uniforms don't show their identities, and the past generations all wore traditional clothes. Why*

---

80 The minorities in China are allowed to have two children. However, the local villagers have strong values that they believe boy is better than girl and can take care of parents when parents get old. During the pregnant period, when it is checked a girl, some of the local villagers choose abortion. Now it is not allowed doctors to tell the gender of the baby before they get birth in China, since the gender problems became an urgent issue. The ratio of men and women in China is imbalanced.

are they now so *renmogouyang*(人模狗样)⁸¹ *hujiahuwei* (狐假虎威),⁸² *and also don't speak Mandarin? Since our indigenous culture is modernized and supposed to serve the tourism economy and make tourists feel more comfortable, they should not just speak a dialect that nobody can understand!*" (Li, 40, female, personal interview, August 6, 2011, 2.3.12)

When talking about the outsiders who work in the village, she said "*Outsiders come here to work but they don't learn and publicize our culture, but despise our culture! When tourist guides bring customers to our village, they should not just bring them to go shopping in shops which are owned by outsiders, and stay in the best hotel in the village. They should introduce them to our indigenous culture and bring them to our historical places. They should not be so material-oriented. Our local village admitted the tourist companies to develop our economy, but the companies should not only think about their business interests. They should consult the local villagers, and deeply understand our local culture. They should not just borrow the modern values and apply to our village, they should listen and learn from the local villagers, and based on the indigenous culture develop a new business strategy, new model and new self-administration method.*" (Li, 40, female, personal interview, August 6, 2011, 2.3.12)

The interview with Li suggests that the local villagers have become slowly aware of their own culture. The tourism economy and diffusion of ICT increased their knowledge of the outside world. In the past, they were less confident about their own culture and attribute the laggard economic status to their indigenous culture. With the tourism economy and diffusion of ICT, many people returned to the villages and the local people get more information about the outside. During this process, they start to reflect and begin rethinking their own culture.

For these places, to deeply understand their own culture, their local government should protect and use modern technologies to record the local culture,⁸³ provide more chances for elderly people to use ICT so that their voices can be

---

81   *Ren* 人 means people, *mou* 模 means shape, *gou* 狗 means dog, *yang* 样 means appearance. This is a four-character idiom in Chinese. The idiom mocks people who look like a dog. Dog in Chinese culture is a belittled word.
82   *Hu* 狐 means fox, *jia* 假 means fake, *hu* 虎 means tiger, *wei* 威 prestige. This is a four-character idiom in Chinese. This idiom means a fox pretends to be a tiger to make a commanding impression.
83   It is already started in North America. There have been a number of oral history projects that focus on using the Internet to capture the thoughts of older people before they pass on, such as American Social History Project – Web Projects, Arizona Memory Project, Birmingham Black Oral History Project, etc. In Germany, the SONIA program aims to bridge the age divide of Internet usage.

heard by society, and in turn the traditional culture can participate in the modern media and become accessible to the young generation. Second, the local village should make their own decisions, and the local government should respect the will of the local villagers. Third, the future of the indigenous culture should be controlled by the local people, not by the tourist companies or agencies.

In the past, increasing numbers of youths have gone to east coast cities to work, as a result, there is a lack of a labor force to develop the local villages. The beautiful landscape and special minority culture cannot keep them in their hometown, but become obstacles and make them lose self-confidence, because for them traditional is equal to bad. Now the tourism economy attracts many tourists, brings local people back to their hometown, and intellectuals, outsiders and people who worked outside bring the new culture to the indigenous place. The new culture is absorbed and integrated into the traditional values. As the modern technology shapes the young generation's values and builds their social reality, the old generation loses their authority, since modern technology endows authority to people. However as the tourism economy continues, the self-realities of individuals are largely changed; at the same time the economy, political engagement, social mobilization, and values are changed as well.

Currently, the TV program called "Jiafeng (家风)" is very popular, jia 家 means home and family, feng 风 means wind, together it means the value or customs of Chinese families. In China, a family is an important unit in traditional culture. The Jiafeng program, by introducing selected families as examples, was created to spread traditional Chinese culture. In the meantime, traditional Chinese food, traditional house styles, traditional clothes and traditional medicine have become popular recently, the media highlighted the values of traditional Chinese culture, and the masses began to appreciate it.

## Conclusion

The minority places are just microcosms to observe the influence of ICT. When a society develops very fast, especially in the field of technology and science, whereas the social system lacks development, the society will be faced with many issues. For instance, the fast development of ICT triggered a trust crisis, individuals' private information was sold, people felt insecure and so on. Elite and well-educated people immigrated to other countries, which not only resulted in the loss of wealth but also brain drain.

The village modernization process is motivated by the tourism economy, which increases the gap between rich and poor and constitutes a 'brain drain' from the minority places toward large urban centers. In the local villages, people who live near the main streets have a better chance to earn money (however, the shops near the street are mostly occupied by outsiders); inside of each village there are many groups, and groups with sightseeing spots can attract more customers; villagers who can use ICT have more opportunity to get tourists (booking, online websites, apps and so on); and villagers who have good connections with tourist agencies always have a better chance to earn money. Local villagers who have money prefer to send their children to study in cities, even though the prosperous local economy attracts outsiders to come and encourages local people to stay, however, most of the young and educated villagers still move out and want to find jobs in the urban areas. These phenomena are epitomes of the Chinese society.

The young generation widely use the new media whereby the old generation cannot use the new media and still depend on the old media to get information—the resulting digital divide in China is also an age divide. The fast diffusion of ICT and the present social memory have started a new chapter, one that is only loosely connected with past chapters. The old generation and young generation mainly represent past social memory and present social memory respectively. The past social memory helps to understand the current social phenomenon, but if it is partly lost, it will be difficult to evaluate the issues of society and predict the future.

A book or a person could be a good analogy to help understand the age digital divide. The continuity of social memory is like a book with contents from the first chapter to the end chapter, or a person who experiences life from an infant period to old age. However, if a book loses many pages at the beginning or a person loses memory, it becomes difficult to predict and understand how the book is going to end and who the person is going to be. The crucial problem is the fragmented Chinese society not only has to face the big gap between rich and poor and imbalanced regional development, but also the difficulty in understanding the current societal situation in order to estimate the future of China.

Countries that are involved in the modernization process are assumed homogenized by the modern culture. However, I found the indigenous place to be not fully eroded by modernization. On the contrary, they partly combined modern values and made the traditional values more adaptable to the social change. The local villagers who are regarded as carriers of the traditional culture adjusted themselves very fast. More and more people are becoming aware of the values of the traditional culture and beginning to appreciate their own culture during the modernization process. There are no linear and direct relationships between modernity and tradition, and traditional culture tends to integrate the modern culture rather than be replaced by it.

# 8 Summary

ICT is expected to hold promise for political engagement, help ordinary citizens to protect their rights, efficiently address economic and social challenges, and provide useful information for agriculture and the medical care of villagers (Albarran et al., 2006; W. L. Bennett, 2008; Hearl, Budge, & Pearson, 1996; Neiger et al., 2012; Norris, 2001; Schroeter, 2012). In the case of China, the information society, the agricultural revolution, and the industrial revolution are all influencing Chinese society at the same time, therefore, ICT in China is struggling to fit into the social change.

The internet arrived in China in 1994. Within two decades, the information industry came to dominate the Chinese economy and changed people's lives dramatically. China, the most rapidly developing and most populous country, has experienced modernization and industrialization at the same time, and is also involved in the wave of globalization and informationization.

The incubator of ICT is the social, political and economic environment of a country when the diffusion processes of innovation begin. The incubator, when ICT started to diffuse in China, was a fragmented social system. The wide gap between rich and poor, eastern and western China, literate and illiterate citizens, urban and rural areas, agricultural and industrial societies, tradition and modernization, etc. have shaped China into a transversally fragmented society where inequality and imbalance are the main characteristics of the society. The reform and opening policy was established in 1978, and since then, the economy is booming. However, behind the rosy picture there are many problems. An investigation titled *The Development and Living Standard of Chinese in 2014* shows that China is becoming the most unequal country in the world. In 1995, the Gini Coefficient was 0.45; by 2002 it had increased to 0.55; however, the figure rocketed to 0.73 in 2012, which suggests that China stands at a crucial juncture, because 1 % of its households own one-third of the property, while 25 % of Chinese households own just 1 % of the property.[84]

The fragmented society of China has a rigid social system, and social facts have become crystallized due to the stability of the planned economy model, as well as the lack of communication between social sectors. After the planned

---

84 See http://society.people.com.cn/n/2014/0725/c1008-25345140.html Accessed August 7, 2015.

economy, China struggled to develop the economy after the Cold War. Deng Xiaoping once described China's process of reform and modernization as being like a person crossing a river by feeling his way over the stones. The central Chinese government is very cautious to every reform. The older social order dominates the society, hindering the formation of a secondary social order. Consequently, the inequality gap will continue to increase instead of shrink, unless a major revolution can break the crystallization—such as the ICT revolution.

The internet gives hope to the younger generation and people of the lower classes, dissolving the rigid stratification of Chinese society. At the same time, people from villages have a chance to work in the city, breaking the *hukou* barrier, which in turn changes the structure of the labor system. The media's job is to foster social cohesion, which helps society integrate the lower classes or those from rural areas. With access to media, they are integrated into modern life and feel that they live together as citizens in the same social system. The development of the Internet accompanied the growth of the new generation, and the new generation found it easier to adapt to the information society than the older generations.

In the polarized, stratified society, the underprivileged were desperate to change their situation, so the whole society welcomed ICT and fully opened the door for them. As a result, ICT has diffused quickly and has changed Chinese society entirely. The society was too starving for a new revolution to break the crystallization, a small push from ICT will bring big changes. The spread of ICT provides a platform that is open for everyone, so it has become a battleground for groups with the same interests. For instance, the popular 50 Cent Party (*wumaodang* 五毛党) and Internet Water Party (*shuijun* 水军) are employed by someone or some groups to write comments and blogs that aim to influence the opinions of the masses. In addition, the public space is rising and the Internet has become a tool for citizens to protect their rights and vent their discontent. Moreover, it also functions as a safety valve, helping society release social stress. It has changed the lives of citizens and also the government's approach to administration.

## The "Social Vacuum System" with Fragmented Social Memory

The communication system of society has been updated since the widespread of ICT. Information is no longer dominated by the government and the masses are not only passive information receivers but also information senders. The top-down communication system has changed to a parallel system, where each person can be both an information producer and receiver.

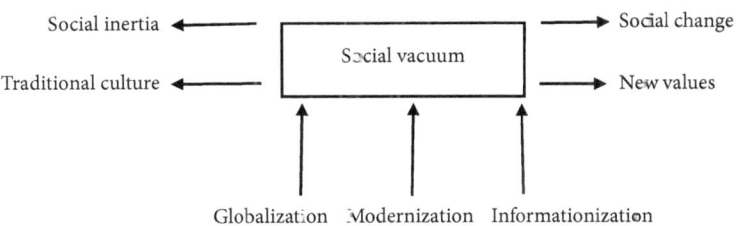

**Fig. 8-1:** Different forces and their impact on the social vacuum state
Source: Author's compilation.

The society has to handle stabilities and changes. On the one hand, the old social system and cultural values like Confucianism, collectivism, and egalitarianism still influence current policies, such as the old planned economy system, the *hukou* policy, the one-child policy, the new socialist villages policy, etc. On the other hand, globalization, modernization, and informationization together with Western culture act like a strong wave impulse on China's socio-economic system as well as the traditional culture. As the deep reform and information revolution progress, there are two forces pulling China: social inertia and fast social change. As a result, the society is formed into a new system, which I call a social vacuum system (see Fig. 8-1).

The "social vacuum system" means the old culture cannot support or cannot meet the requirements of the transition. Meanwhile, new values exist as an immature culture that struggles to form mainstream values to support social activities. Faced with social change, the traditional cultures cannot provide or give constructive suggestions for action to individuals who live in this system. Their values are more formed by current society and both traditional culture and current values fail to guide the actors. If social memory is regarded as a book, where the memories and knowledge of each person are part of the book, if some people lack access to the means of expressing and recording their knowledge, the whole book will not continuously develop.

What is the influence of ICT on the fragmented society? Firstly, one of the functions of ICT is the redistribution of resources. China has jurisdiction over 22 provinces, five autonomous regions, four directly controlled municipalities (Beijing, Tianjin, Shanghai, and Chongqing), and two mostly self-governing special administrative regions (Hong Kong and Macau). There are major differences between the regions and provinces. Earlier policies, such as *hukou* and the planned economy, have divided China into a dual urban and rural society.

However, in the information society, ICT skills are important for finding work. Hence, the geographic differences are not a barrier between rich and poor, but the command of the Internet.

The diffusion of ICT provides opportunities for people who live in rural areas and have been kept out of the cities by the *hukou* policies. Nowadays, most of the workers in China are young people who can positively interact with ICT and can easily accept the new technologies. ICT stimulates social mobility by opening a door for people to climb to the upper classes and improve their social status.

Secondly, the new ICT, especially the Internet provides a safety valve for society, a sort of psychological equilibrium for people who live in the rural areas. The Internet is a public resource open to everyone, so users receive and also send information. People who live in the rural areas can be connected by ICT, which increases social cohesion and gives them a feeling that they belong to the larger society.

Unlike Western culture, where the individual is seen as a unit of society, in China the family is regarded as the basic unit of society. Thus, if one person in a family can use the Internet, then he or she will help other family members use it.[85] If the rights of one member of a family are infringed, the family can probably mobilize more relatives to protect his or her rights. The Internet plays a major role in this process, as the more internet-savvy family members will use it to publicize the infringement, hoping to a create scandal.

Thirdly, the hidden contradictions will be exposed to the sun while widespread usage of the Internet leads to the unprecedented expansion of information. Before the Internet was exported to China, information was occluded by the central and local governments. The process of information transfer is: first, the central government sends the documents to the local government; afterwards the local government sends the news to the civil servants. It is worth noting that government news also has a hierarchy, and that certain levels of the local government can only get information according to their level. This means that the government of a town never gets the same information as the government of a province. Similarly, the officers cannot get the same information as their leader. For the ordinary citizens, political information is hard to get. However, ICT has

---

85 See the case of Pianpo, Chen as an intellectual of the village who cannot use a computer. However, his son can, so he can help him check information online. As is traditional in Chinese culture, elderly people are always cared for by their family instead of being sent to care facilities. The old generation cannot use the Internet, but the young generation can help them.

opened a window to the public, and the information hierarchy is gradually vanishing—ordinary ICT users can get the same information as people who work for the government.

At the same time, the diffusion of ICT also brings instability to Chinese society. For example, terrorist attacks in Xinjiang and Tibet have aroused debates in the whole country. As for minority issues, the conflict between minority citizens and their local governments appears to have become acute in recent years. The government used to hide conflicts and journalists were forbidden to release news that might cause social disorder. However, the media now exposes more and more conflicts, leading to public anxiety that fewer tourists will want to visit minority areas. But the truth is that the conflicts have existed for a long time, and the public only believes that the situation is becoming worse because the news is now available on the Internet, and the messages are difficult to control. Meanwhile, the media publishes details of crimes, many potential criminals refer to them, and the crime rate increases. When the first violent attack happened in the kindergarten, the criminal used a knife and killed many children. The whole country was shocked. The media published details about how the criminal prepared and conducted his plan, and this information was then utilized by copycats who used the same methods to attack other kindergartens. Thus, crimes replicate themselves via the media.

Fourthly, social cohesion is increased with the spread of ICT. Whether in advertisements or in TV dramas, the media try their best to show that people are living in a prosperous age. Especially when festivals are nearby, their warm and happy atmosphere is exaggerated. Such shows always use big cities as the background, so that when the audience watches TV, they feel connected with the whole society, and ignore the reality of their own situation. In this way, the media function like a bridge that connects real life and comfortable illusion. People living in rural areas do not belong to the big cities, but psychologically, with the help of the media, they feel that they live in the same world as their urban counterparts, even though the gap between them is huge.

Last but not least, the spread of the media accelerates knowledge transfer. For instance, there has been a general increase in legal knowledge since the government propagates the notion that China is a free society ruled by law, the media have an obligation to educate the public on legal matters. The masses get legal knowledge from the media and then use the law to protect themselves when their rights are violated.

However, as the ICT diffuses broadly, the society starts to become a fragmented society, as the society breaks with the earlier generations and the social memory lose continuity. In the Micro Narrative period, information from the Internet and news from the media tend to get shorter and fragmented. The news prefers to use shrilling, sensational and interesting headlines to attract audiences. In China, the young generation uses the new ICT to record, work, and communicate, and they record almost everything in the Internet, which constitutes the most important part of the present social memory. Moreover, economic and social activities are closely linked to the Internet as well, and the industrial revolution is now being updated to use computer control systems in order to enter into the smart era (Industry 4.0). The rapid and extreme change will cause many challenges for the older generations as well as the less educated people. Most of the older generations in China could not use smartphones to book a hotel, order a taxi, order food, or shop online. The older generations are losing their voice in society and being forced to get off the "fast speed train".

One of the reasons which contribute to the old generation having difficulty adjusting to the information society is the change of language system. After China was established in 1949, the government simplified the Chinese characters and referred to the Roman alphabet to invent Pinyin for writing and pronouncing Chinese. The new digital devices are mostly based on the Pinyin system to type. The older generations who are less educated, and cannot write simplified Chinese and Pinyin are hardly able to use ICT to spread their ideas and knowledge. In the information age, the old generations are passive information receivers and they prefer to use old media. The problem is the traditional media and new media have sharp differences in content that can directly shape users' values and social realities. Therefore, the young generation as the main carrier of modernization, and the old generation as the carriers of tradition, might have different social realities. As a result, the modern social memory might fail to connect with the old social memory, which might bring confusion to individuals, who then lose their identities.

The age divide is also a problem in the Western countries. However, it is worth noting that the modernization of indigenous cultures has developed over centuries and that society develops step by step. In the historical perspective, when an agricultural society changes into an industrial society, the big social transition may face a similar situation; that the old social system and culture lag behind the social change. This can be explained by tracing back to when the industrial revolution first started in the West. When industrialization started, the agricultural society might have failed to meet the new changes of the industrial revolution, and it took time for the social system to adjust to the social change. Then, step by step, the West faced the information revolution, and the society took decades to finally form a stable social system. Whereas Western countries had centuries to reach a stable state with a revised social structure compatible with modernization, China has undertaken three revolutions (agricultural, industrial, information) in a much shorter span of time.

A balanced development of society does not mean that the technologies develop ahead of the other aspects; the social adjustment and psychological adaption of individuals need more time than the adoption of new technology. When a society moves from the agricultural revolution to the industrial revolution, there is a time lag as individuals and social systems adjust to a new era. However, in China, the three revolutions overlap, and the third revolution developed particularly rapidly, which changes the implements of narration and affects the social memory directly.

Touraine describes the society of France as changing from a pyramid structure to a marathon. The pyramid structure may have different hierarchies, but at least different classes stay in the same structure. However, the marathon is different, as some participants are dropping behind every second. The one who falls behind is unlikely to catch up to the team, will be abandoned, and may never join the race again. At the moment, in France, some workers are still keeping pace with the game, but most of them have given up. The participants who are still running are the employees absorbed by the global economic system (Sun, 2004).

In China, the situation is similar, as Touraine has claimed that there are many people lagging behind in the race and abandoned by society. There are four groups among the laggards. The first is unemployed workers. At the end of the 1990s, the reform of state-owned companies meant that many workers were laid-off because there were not enough jobs. Most of the laid-off workers were uneducated and unskilled. The second group is the illiterate and disadvantaged people. As labor-intensive industries are replaced by technology-intensive industries, many companies prefer workers with a good education. The third group is the minorities who live in a rural area and cannot speak Mandarin. In western China, transportation is still the biggest problem, even though the situation is improving, but the minority regions are relatively poor compared to other parts of China. The fourth group is the people who lack access to information. Most of these are elderly people who cannot use the Internet.

According to the Dependency Model of Mass Media, the media can influence an individual's values if he or she does not have strong faith, values, and sense of self. News is selected by the media, which ultimately influences users' values. Different media select different news. When audiences choose television for access to information, the information they get is more homogeneous than that found in new media like smartphones or computers. The content of the Internet is 'darker' than the rosy scenario on television and it is becoming a means to expose the dark side of society, such as political scandals. Thus, different media users will have different views toward society.

The biggest consequence of a fragmented society is that the diffusion of ICT breaks the social memory into two parts: past social memory and present social memory. The modern society lacks a connection to the traditional culture and traditional culture loses the ability to support the new system. Modernity as the dominant culture influences society, and there is less communication with traditional society. Thus, Chinese society will be like the indigenous villages in that as the traditional culture is replaced by modern culture, the Chinese society will face brain drain and a sharp gap between rich and poor, and the people who are left will question their identities and feel lost in the modern world.

Currently, one important topic on the Internet is whether Lei Feng existed or not. President Xi Jinping has started to attach importance to Lei Feng's spirit. However, many young and middle-aged people doubt that Lei Feng ever existed, suspecting he was a fictitious person created to serve as a Maoist moral model. Since the young generation has not experienced Lei Feng's period, no one knows whether he is an imaginary figure or not. In Mao's period, Lei Feng[86] was a famous moral model for the masses and the central government encouraged them to learn from him. The elderly generation is familiar with Lei Feng, and some of them worked with him, but the debate excludes them and they cannot participate in the discussion because of their lack of computer skills. If the elderly people do not clarify the truth, it will be buried by the Internet.

ICT is an important way to record history. As the computer has become a major tool for work and recording, elderly people cannot use the Internet or computers so they are silenced. Therefore, the elderly who are willing to use the new ICT are the most important part of the information society, as they are linking modernity and tradition. If society is a book, it is difficult to understand if the reader skips the beginning chapters, or loses some of the contents. When we view the society in this perspective, excluding the elderly from the information economy means we lose some part of the social memory, as they are the continuity of the whole system. In our book metaphor, the elderly is the introduction and first chapters of the books. They represent the traditional cultures and they are an important force to bridge traditional culture and modern culture.

---

86  Lei Feng was born on 18 December 1940 in Changsha and died on 15 August 1962. He was a soldier and was characterized as a selfless and humble person who dedicated his life to the Communist Party. Mao Tse-tung encouraged the whole country to learn from him.

Developing countries can borrow technologies directly from developed countries, and innovation takes less time to research and develop. However, a balanced development of society will not necessarily follow. The basic difference between the industrial revolution and the information revolution is that the former changes the implements of production and the latter changes the implements of narration.

ICT is an important tool to shape the social memory. As ICT spreads, society faces not only imbalanced development but also a social memory break. The penetration of ICT tends to melt the crystallized society and shrink the gap between rich and poor, decrease the divide among different regions and so on. However, in the long term, the fragmented society has created the present social memory disjunction, and that will become the biggest problem. China as known, with its long history deeply influenced by Confucianism, may not exist. Prior knowledge cannot be used to understand the new China, since current China is shaped by ICT and the present social memory has started a new chapter that might close the connection with past chapters.

# List of Figures

| | | |
|---|---|---|
| Fig. 2-1: | Knowledge transfer in authoritarian theory adapted from the four theories of the press | 32 |
| Fig. 3-1: | Media and social condition model | 49 |
| Fig. 3-2: | Annual average number of newspaper and magazine subscribers per 100 persons | 50 |
| Fig. 3-3: | Adoption number of television and radio from 1982 to 1984 | 51 |
| Fig. 3-4: | News transmission in Mao's period | 53 |
| Fig. 3-5: | Large newspaper groups in China | 64 |
| Fig. 3-6: | The telecommunications reforms in China | 79 |
| Fig. 3-7: | Internet penetration and participation of the top ten languages used in the Web in 2015 | 89 |
| Fig. 3-8: | Users of fixed and mobile phones from 1998 to 2012 | 93 |
| Fig. 4-1: | The trend of fixed-phones in China from 2002 to 2015 | 98 |
| Fig. 4-2: | The trend of mobile phones in China from 2002 to 2016 | 99 |
| Fig. 4-3: | The fixed-phone diffusion rate of each province from 2002 to 2016 | 100 |
| Fig. 4-4: | The mobile phone diffusion rate of each province from 2002 to 2016 | 101 |
| Fig. 4-5: | The diffusion of the internet in China from 2003 to 2013 | 102 |
| Fig. 4-6: | The trend of fixed-phone diffusion between 2000–2014 | 104 |
| Fig. 4-7: | The fixed-phone diffusion rate from 2000 to 2014 | 106 |
| Fig. 4-8: | The trend of smartphone diffusion between 2000–2014 | 107 |
| Fig. 4-9: | The smartphone diffusion rate from 2000 to 2014 | 108 |
| Fig. 4-10: | The trend of broadband diffusion between 2000–2014 | 110 |
| Fig. 4-11: | The broadband diffusion rate from 2000 to 2014 | 111 |
| Fig. 4-12: | The diffusion curves of innovations | 112 |
| Fig. 5-1: | The relationship between the rate of public attention to Guo Meimei and the donation amount | 123 |
| Fig. 6-1: | Location of the sampling area | 130 |
| Fig. 6-2: | Population distribution of economic sectors of the Qiandongnan Miao and Dong Autonomous Prefecture | 134 |
| Fig. 6-3: | The diffusion of technologies in Dong Autonomous Prefecture from 2011 to 2015 | 139 |
| Fig. 7-1: | Reasons to not use the Internet | 173 |
| Fig. 7-2: | Age of Internet users | 174 |
| Fig. 8-1: | Different forces and their impact on the social vacuum state | 185 |

# List of Tables

| | | |
|---|---|---|
| Tab. 3-1: | Authors' Calculations based on CNNIC Statistics in 2000 and 2002 | 77 |
| Tab. 6-1: | Revenue resources of the investigated region in 2011 and 2015 | 135 |
| Tab. 6-2: | Comparison of information access in 2010 and 2015 | 143 |
| Tab. 6-3: | Telephone number attitude in 2010 and 2015 | 148 |
| Tab. 6-4: | Entertainment activities before and after tourism | 153 |
| Tab. 6-5: | Most Favorite TV programs | 155 |
| Tab. 6-6: | The frequency of wearing traditional clothes | 158 |
| Tab. 6-7: | People to ask for help when conflict happens | 160 |
| Tab. 6-8: | The opinion of tourism (multiple choices). | 160 |
| Tab. 6-9: | Opinion of outsiders who work in the villages | 160 |
| Tab. 7-1: | How local villagers feel about showing their traditional activities to tourists. | 177 |

# Bibliography

Adams, V. (1992). Tourism and Sherpas, Nepal: reconstruction of reciprocity. Annals of tourism research, 19(3), 534–554.

Adoni, H., & Mane, S. (1984). Media and the social construction of reality toward an integration of theory and research. Communication research, 11(3), 323–340.

Ai, H. (2002). Zhongguo guangbo dianshishi chulun [Introduction of Broadcast History in China]. Jinan: Shandong daxue chubanshe.

Albarran, A. B., Chan-Olmsted, S. M., & Wirth, M. O. (2006). Handbook of media management and economics. London: Routledge.

Alitto, G. (1986). The last Confucian: Liang Shu-ming and the Chinese dilemma of modernity. Berkeley: University of California Press.

Ames, R., & Hall, D. (2015). The democracy of the dead: Dewey, Confucius, and the hope for democracy in China Illinois: Open Court.

Angell, I. (1996). Winners and losers in the information age. Society, 34(1), 81–85.

Ansolabehere, S., & Warburn, P. (1995). Evolving perspectives on the effects of campaign communication. Research in Political Sociology, 7-8, JAI.

Antonelli, C. (2003). The digital divide: understanding the economics of new information and communication technology in the global economy. Information Economics and Policy, 15(2), 173–199.

Ao, Y. (2016). Fu Sinian weihe tan ziji yishiwucheng [Why Fu Sinian thinks he achieve nothing]. Retrieved June 8, 2016 from http://xw.cq.com/cul/20151011010345/CUL2015101101034503.

Ausubel, J. H. (1991). Rat Race Dynamics and Crazy Companies: The Diffusion of Technologies and Social Behavior Diffusion of Technologies and Social Behavior. In Nakicenovic, N., & Grübler, A. (Eds.), Diffusion of Technologies and Social Behavior (pp. 1–17). Heidelberg: Springer.

Babel, & W3Techs. (2015). Content languages for websites. Retrieved May 6, 2015, from https://w3techs.com/technologies/overview/content_language/all.

Ball-Rokeach, S. J., & DeFleur, M. L. (1976). A dependency model of mass-media effects. Communication Research, 3(1), 3–21.

Bandura, A. (2001). Social cognitive theory of mass communication. Media psychology, 3(3), 265–299.

*Barzilai-Nahon, K., Rafaeli, S., & Ahituv, N.* (2004). Measuring gaps in cyberspace: Constructing a comprehensive digital divide index. Paper presented at the Workshop on Measuring the Information Society, the conference of Internet Research, 2015 November 30; Hiroshima, Japan.

*Basu, P., & Chakraborty, J.* (2011). New technologies, old divides: linking internet access to social and locational characteristics of US farms. GeoJournal, 76(5), 469–481.

*Bauman, Z.* (1997). Hermeneutics and modern social theory. Anthony Giddens: Critical Assessments, 1, 211.

*Baumol, W. J.* (1986). Entrepreneurship and a century of growth. Journal of Business Venturing, 1(2), 141–145.

*Beck, U.* (1992). Risk society: towards a new modernity (Vol. 17). London: Sage.

*Bell, D. A.* (2010). China's new confucianism: politics and everyday life in a changing society. Princeton: Princeton University Press.

*Bennett, C. J., & Howlett, M.* (1992). The lessons of learning: reconciling theories of policy learning and policy change. Policy sciences, 25(3), 275–294.

*Bennett, W. L.* (1990). Toward a theory of press-state relations in the United States. Journal of communication, 40(2), 103–127.

*Bennett, W. L.* (2008). Changing citizenship in the digital age. Civic life online: learning how digital media can engage youth, 1, 1–24.

*Benton, G.* (1984). Chinese communism and democracy. New left review, 148, 57.

*Biggerstaff, K.* (1966). Modernization and early modern China. The journal of Asian studies, 25(04), 607–619.

*Birner, J., & Ege, R.* (1999). Two views on social stability: an unsettled question. American journal of economics and sociology, 58(4), 749–780.

*Bittner, J. R.* (1977). Mass communication, an introduction: theory and practice of mass media in society. Mahwah: Lawrence Erlbaum Associates.

*Black, C. E.* (1975). The modernization of Japan and Russia: a comparative study. New York: Free Press.

*Black, C. E.* (1976). Comparative modernization: a reader (Vol. 1). New York: Free Press.

*Blanchard, M. A.* (2013). History of the mass media in the United States: an encyclopedia. London: Routledge.

*Bloch, A.* (2003). Murphy's law. London: Penguin.

*Bloom, D. E., Canning, D., & Fink, G.* (2008). Urbanization and the wealth of nations. Science, 319(5864), 772–775.

*Blumler, J. G., Blumler, J.*, & *Gurevitch, M.* (1995). The crisis of public communication. London: Psychology Press.

*Bo, Y.* (1986). Chouloude zhongguoren [The ugly Chinese]. Guangzhou: Huacheng chubanshe.

*Bosch, F.* (2015). Mass media and historical change: Germany in international perspective, 1400 to the present. Oxford: Berghahn Books.

*Bottero, W.* (2004). Stratification: social division and inequality. London: Routledge.

*Brandtzæg, P. B., Heim, J.*, & *Karahasonović, A.* (2011). Understanding the new digital divide—a typology of internet users in Europe. International journal of human-computer studies, 69(3), 123–138.

Broadcast. (2013). Guangbo dianshi fazhan lishi [The History of Broadcast and Televison Development]. Retrieved June 7, 2016, http://zhishi.maigoo.com/85621.html.

*Brown, B. E.* (1969). The French Experience of Modernization. World Politics, 21(03), 366–391.

*Cai.* (2003). Guanzhide huanghui:zhongguo dianxinye wanyiyuan chongzu shilun [Sunset of control: the record of telecommunication reform in China]. Shanghai: Shehui kexue wenxian chubanshe.

*Carey, J. W.* (1993). The mass media and democracy: between the modern and the postmodern. Journal of international affairs, 1, 21.

*Chang, H.* (1976). New confucianism and the intellectual crisis of contemporary China. Cambridge: Harvard University Press.

*Chao, L.*, & *Myers, R. H.* (1998). The first Chinese democracy: political life in the republic of China on Taiwan. Maryland: Johns Hopkins University Press.

*Chen, M., Liu, W.*, & *Tao, X.* (2013). Evolution and assessment on China's urbanization 1960–2010: under-urbanization or over-urbanization? Habitat international, 38, 25–33.

*Chen, X.* (2010). Xiangcun zhili jingying zhuanxing wenti tantao [Transform of elite and development of village]. Zhongguo zhengzhi, 2-8.

*Cheng, C.-Y.*, & *Bunnin, N.* (2008). Contemporary Chinese philosophy. New York: John Wiley & Sons.

*Chotpitayasunondh, V.*, & *Douglas, K. M.* (2016). How "phubbing" becomes the norm: The antecedents and consequences of snubbing via smartphone. Computers in human behavior, 63, 9–18.

*Chyi, H. I.*, & *McCombs, M.* (2004). Media salience and the process of framing: Coverage of the Columbine school shootings. Journalism & mass communication quarterly, 81(1), 22–35.

*Clark, P.* (1987). Chinese cinema: culture and politics since 1949. Cambridge, UK: CUP Archive.

*CNNIC.* (2000). "Diwuci hulianwang fazhan diaocha baogao [CNNIC annual report 2000]. Retrieved October 16, 2015, from https://www.cnnic.net.cn/hlwfzyj/hlwxzbg/index_5.htm.

*CNNIC.* (2007). "2007nian zhongguo nongcun hulianwang diaocha baogao 2007 [ The investigation of internet usage in rural China in 2007]. Retrieved June 29, 2015, from http://www.cnnic.net.cn/hlwfzyj/hlwxzbg/ncbg/201206/t20120612_27435.htm.

*CNNIC.* (2016). CNNIC 37th report. Retrieved October 26, 2016, from http://www.cac.gov.cn/2016-01/22/c_1117858695.htm.

*CNNIC.* (2017). Reports of CNNIC. Retrieved February 6, 2017, from https://www.cnnic.net.cn/hlwfzyj/hlwxzbg/.

*Commerce, U. S. D. o., & Brown, R. H.* (1995). Falling through the net: a survey of the "Have nots" in rural and urban America. Washington, D.C: National Telecommunications and Information Administration • U.S. DEPARTMENT OF COMMERCE.

*Comte, A.* (1856). Social physics: from the positive philosophy. New York: Calvin Blanchard.

*Confucius.* (2012). Lunyu XII. Retrieved December 11, 2014, from http://wengu.tartarie.com/wg/wengu.php?l=Lunyu&no=312.

*Cooley, C. H.* (1909). Two major works: social organization. Human nature and the social order. Glencoe: Free Press.

*Couldry, N.* (2012). Media, society, world: Social theory and digital media practice. Cambridge: Polity.

*Crenshaw, E. M., & Robison, K. K.* (2006). Globalization and the digital divide: the roles of structural conduciveness and global connection in internet diffusion*. Social science quarterly, 87(1), 190–207.

*Crook, S., Pakulski, J., & Waters, M.* (1992). Postmodernization change in advanced society. London: Sage.

*Cui, Y.* (2009). "ziyoou"de shenghuo neihan he lilun neihan [The literature and real meaning of freedom]. shanghai shifan daxue xuebao shehui kexueban, 38(4), 19–27.

*Davison, R. M.* (2016). Review of: ICTs in developing countries: research, practices and policy implications. The electronic journal of information systems in developing countries, 73-75.

*DeFleur, M. L.* (1966). Theories of mass communication. Pennsylvania: McKay.

DiMaggio, P., & Hargittai, E. (2001). From the 'digital divide' to 'digital inequality': Studying Internet use as penetration increases. Princeton University, 4(1), 4–2.

DiMaggio, P., Hargittai, E., Celeste. C., & Shafer, S. (2004). From unequal access to differentiated use: A literature review and agenda for research on digital inequality. Social inequality. Retrieved August 19, 2016, from http://www.russellsage.org/sites/all/files/u4/DiMaggio%20et%20al.pdf

Ding, W. (1935). Zailun minzhu yu ducai [Re-discuss about liberty and autarchy]. Tianjin: Dagongbao.

Dolowitz, D., & Marsh, D. (1996). Who learns what from whom: a review of the policy transfer literature. Political studies, 44(2), 343–357.

Dolowitz, D. P. (2000). Policy transfer and British social policy: learning from the USA?. London: Open University Press.

Du, J. (2003). The spread of gossip: A mass media analysis of the spread of SARS-related news. Journal of Nanjing University (philosophy, Humanities and social sciences), 40, 155.

Durkheim, E. (1933). The division of labor in society. Translated by Halls, W. D. New York: Free Press.

Eickelman, D. F., & Anderson, J. W. (2003). New media in the Muslim world: the emerging public sphere. Bloomington: Indiana University Press.

Epstein, I. (1982). Educational television in the People's Republic of China: some preliminary observations. Comparative education review, 26(2), 286–291.

ESDQMDAP. (2016). Data of economic and social development of Qiandongnan Miao and Dong autonomous prefecture. Retrieved August 18, 2016, from http://www.qdn.gov.cn/xxgk/zdgk/tjxx/tjnb/.

Fang, H. (1981). Zhongguo jindai baokanshi [The press history of early China]. Taiyuan: Shanxi jiaoyu chubanshe.

Fang, I. E. (1997). A history of mass communication. Boston: Focal Press.

Fei, X. (1983). Rural development in China: Prospect and retrospect. Chicago: University of Chicago Press.

Fei, X. t. (1997). Fansi, duihua, wenhuazijue [Reflection, communicate, cultural awareness]. Beijing daxue xuebao (shehui kexueban), 3–5.

Feng, X. (2000). Dousing zinv qingshaonian de Chihuahua jointing jiqi jieguo [The socialization procedure and consequences of children under One Child Policy]. zhongguo shehui kexue, (6), 118–131.

Fixed-phone. (2012). Duding dianhua fazhanshi [The history of fixed-phone]. Retrieved July 6, 2016, from http://wenku.baidu.com/view/6e97017f1711cc7931b71682.html

*Foucault, M., Morris, M., & Patton, P.* (1979). Michel Foucault power, truth, strategy. New York: Prometheus Books.
*Freeman, L. A.* (2000). Closing the shop: information cartels and Japan's mass media. Princeton: Princeton University Press.
*Friedrich, L. K., & Stein, A. H.* (1975). Prosocial television and young children: the effects of verbal labeling and role playing on learning and behavior. Child development, 46 (1), 27–38.
*Gao, D.* (2000). The intellectual's voice. Beijing: Beijing shiyue wenyi chubanshe.
*Gao, X.* (2008). Zhongguo dianxin shichangde qu longduan gaige yu jishu jinbu [The anti-monopoly of telecommunication reform and technological progress of China]. Jingji kexue, 6, 66–77.
*Gilboa, I., & Matsui, A.* (1991). Social stability and equilibrium. Econometrica: journal of the econometric society, 859–867.
*Gitlin, T.* (1980). The whole world is watching: mass media in the making & unmaking of the new left. Oakland: University of California Press.
*Glauser, B.* (2015). Street children. Constructing and reconstructing childhood: contemporary issues in the sociological study of childhood. Washington: Falmer Press.
*Glaziev, S.* (1991). Transformation of the Soviet economy: economic reforms and structural crisis. National Institute Economic Review, 138(1), 97–108.
*Gorman, L., & McLean, D.* (2002). Media and society in the twentieth century: a historical introduction. Hoboken: Wiley-Blackwell.
*Gough, H.* (1988). The newspaper press in the French Revolution. London: Taylor & Francis.
*Graber, D. A.* (2009). Mass media and American politics. New York: Sage.
*Grieder, J. B.* (1970). Hu Shih and the Chinese renaissance: liberalism in the Chinese Revolution, 1917-1937 (Vol. 46). Cambridge: Harvard University Press.
*Grin, J., & Loeber, A.* (2006). Theories of policy learning: agency, structure, and change. Handbook of public policy analysis (pp:201–219). New York: CRC Press.
*Grübler, A.* (1991). Diffusion of technologies and social behavior. Heidelberg: Springer.
*Gu, C.* (2004). Gaige knifing yilai zhongguo chengshihua yu jingji shehui fazhan guano yanjiu [The relationship between urbanization and development of socio-economic development after Chinese economic reform]. Renwen Dili, 19(2), 1–5.

*Gui, Y., & Xu, J.* (2012). The butterfly effect invoked by personal information----a case of Guo Meimei in Weibo Age. Digital strategy, 2-4.

*Gunter, B.* (1987). Poor reception: misunderstanding and forgetting broadcast news. London: Routledge.

*Guo, M.* (1928). Zhuozi shangde tiaowu [Dance on the table]. Chuangzao yuekan, 1, 11.

*Haddon, L.* (2004). Information and communication technologies in everyday life: a concise introduction and research guide (new technologies/new cultures). Oxford: Berg.

*Hall, P. A.* (1993). Policy paradigms, social learning, and the state: the case of economic policymaking in Britain (pp: 275–296). Comparative politics. New York: University of New York.

*Hallin, D. C., & Mancini, P.* (2011). Comparing media systems beyond the Western world. Cambridge: Cambridge University Press.

*Hao, Y., & Lu, H.* (2012). Wangmin jiti xingdongde dongli jizhi tanxi----yi "Guo Meimei shijian" wei yanjiu ge'an [The mechanism of Netizen s motivation----a case study of Guo Meimei]. Guoji xinwenjie, 34(3), 61–66.

*Hargittai, E.* (2002). Beyond logs and surveys: In-depth measures of people's web use skills. Journal of the American society for information science and technology, 53(14), 1239–1244.

*Hargittai, E.* (2005). Survey measures of web-oriented digital literacy. Social science computer review, 23(3), 371–379.

*Hargittai, E.* (2009). An update on survey measures of web-oriented digital literacy. Social acience computer review, 27(1), 130–137.

*He, X.* (2014). Woguo baozhi gongnengde yanli yanjiu [The function of newspaper in a historical approach]. Chunbo yu banquan, 3, 3–4.

*Hearl, D. J., Budge, I., & Pearson, B.* (1996). Distinctiveness of regional voting: a comparative analysis across the European Community (1979–1993). Electoral studies, 15(2), 167–182.

*Heeks, R.* (2002). i-development not e-development: Special issue on ICTs and development. Journal of international development, 14(1), 1–11.

*Held, D.* (1995). Democracy and the global order: From the Modern State to Cosmopolitan Governance. California: Stanford University Press.

*Herman, E. S., & Chomsky, N.* (2010). Manufacturing consent: The political economy of the mass media. New York: Random House.

*History.* (2011). Jindai zhongguo dianhua fazhanshi [The fixed-phone development history of early China]. Retrieved May 21, 2014, from http://wenku.baidu.com/view/2a6c7cb269dc5022aaea009d.html.

*History*. (2012). "Zhongguo dianshiji fazhan licheng [History of China's televison]. Retrieved May 21, 2014, from http://wenku.baidu.com/view/b2fe98c58bd63186bcebbc39.html.

Hocking, W. E. (1948). Freedom of the press. The philosophical review, 57 (2). 186–190.

Hoffman, D. L., Novak, T. P., & Schlosser, A. E. (2001). The evolution of the digital divide: examining the relationship of race to Internet access and usage over time. In Compaine B. M. (Ed.), The digital divide: facing a crisis or creating a myth, (pp. 47–97). Cambridge: MIT Press.

Horowitz, I. L. (1976). Personality and Structural Dimensions in Comparative International Development. In C. E. Black (Ed.), Comparative Modernization: A Reader, (pp. 494-513). Washington, D.C: Free Press.

Hsu, L. S.-l., & Sun, Z. (1933). Sun Yat-sen, his political and social ideals: a source book. Oakland: University of Southern California Press.

Hu, S. (1998). Renquanlun ji [Human right]. Beijing: Beijing daxue chubanshe.

Hu, S. (2013). wode qilu [The wrong road of mine]. Beijing: ershiyi shiji chubanshe.

Huang, K.-w. (2003). The reception of Yan Fu in twentieth-century of China. Lanham: University Press of America.

Huang, K. (2008). The meaning of freedom: Yan Fu and the origins of Chinese liberalism. Hong Kong: Chinese University Press Hong Kong.

Humphreys, P. (1996). Mass media and media policy in Western Europe (Vol. 2). Manchester: Manchester University Press.

Huntington, S. P. (1971). The change to change: modernization, development, and politics. Comparative politics, 3(3) (Apr., 1971), pp. 283–322.

Inglehart, R. (1997). Modernization and postmodernization: cultural, economic, and political change in 43 societies (Vol. 19). Cambridge: Cambridge University Press.

Inglis, F. (1990). Media theory: an introduction. Oxford: B. Blackwell.

Inkeles, A. (1976). A Model of the Modern Man. In Black, C. E. (Ed.), Comparative Modernization: A Reader. Washington, D.C: Free Press.

*Internet*. (2004). Zhongguo hulianwang fazhan dashiji [Record of big events of China's internet]. Retrieved May 28, 2015, from http://tech.sina.com.cn/i/2005-07-19/1038666886.shtml.

*Internet*. (2012). Zhongguo hulianwang fazhan de sige jieduan [Four stages of internet development of China]. Retrieved March 15, 2015, from http://www.360doc.com/content/12/0223/22/153944_189135292.shtml.

*Internet*. (2014). Zhongguo hulianwang [The internet of China]. Retrieved March 15, 2015, from http //news.xinhuanet.com/ziliao/2003-01/22/content_702667.htm.

Ip, H.-Y. (1994). The origins of Chinese communism: a new interpretation. Modern China, 20(1), 34–63.

*ITU*. (2015). "Database of ITU". Retrieved June 6, 2015, from http://www.itu.int/en/ITU-D/Statistics/Pages/default.aspx.

*IWS*. (2015a). "Internet penetration and participation of the top ten languages used in the Web in 2015". Retrieved March 21, 2015, from http://www.internetworldstats.com/stats7.htm

*IWS*. (2015b). "The top ten languages used in the web in 2015". Retrieved March 21, 2015, from http://www.internetworldstats.com/stats7.htm.

Jackson, L. A., Zhao, Y., Qiu, W., Koleric III, A., Fitzgerald, H. E., Harold, R., & Von Eye, A. (2008). Cultural differences in morality in the real and virtual worlds: a comparison of Chinese and US youth. CyberPsychology & behavior, 11(3), 279–286.

Jiang, L. (2004). Nongmingong zai chengshide shengcun yu shiying [Investigation of rural villagers work in cities---the living and modernity of rural villagers]. Zhengzhou daxue xuebao: zhexue shehui kexueban, 37(1), 74–74.

Jin, M. (2003). feidian" shiqi meitide zeren [Responsibilities of media during SARS period]. zhonggong ningbo shiwei dangxiao xuebao, 25(3), 79–80.

Judge, J. (1996). Print and politics: 'Shibao' and the culture of reform in late Qing China. Stanford: Stanford University Press.

Jung, H. Y. (2002). Comparative political culture in the age of globalization: an introductory anthology Lanham: Lexington Books.

Jussawalla, M. (1999). The impact of ICT convergence on development in the Asian region. Telecommunications policy, 23(3), 217–234.

Kamarunzaman, N. Z., Zakaria, Z., Zawawi, A. A., Noordin, N., Sawal, M. Z. H. M., Rahman, A. E. A., Norain, M. M. B. (2011). Identifying gaps in digital divide-comparison between localities in Sg. Petani, Kedah, Malaysia. Interdisciplinary journal of contemporary research in business, 3(4), 556.

Kassarjian, H. H. (1965). Social character and differential preference for mass communication. Journal of marketing research, 2(2) (May, 1965), pp. 146–153.

Katz, E., Haas, H., & Gurevitch, M. (1973). On the use of the mass media for important things. American sociological review, 38 (2), 164–181.

*Katz, E., & Lazarsfeld, P. F.* (1955). Personal influence, the part played by people in the flow of mass communications. Piscataway: Transaction Publishers.

*Keane, J.* (1998). The media and democracy. Cambridge: Polity Press.

*Keller, W.* (2001). International technology diffusion. Retrieved July 18, 2016, from http://spot.colorado.edu/~kellerw/ITD.pdf.

*Kerbel, M. R.* (1999). Remote & controlled: media politics in a cynical age. Boulder: Westview Press.

*Kerr, C.* (2001). The uses of the university. Cambridge: Harvard University Press.

*Kleinnijenhuis, J.* (1991). Newspaper complexity and the knowledge gap. European journal of communication, 6(4), 499–522.

*Kohn, L.* (2001). Daoism and Chinese culture. Dunedin: Three Pines Press.

*Kohut, H.* (1976). Creativeness, charisma, group psychology. The search for the self, 2, 287–301.

*Kong, G., & Jin, Z.* (2003). Xiaolingtong dui zhongguo tongxin shichang de yingxiang [The influence of Xiaolingtong on the telecommunication market]. Guanli shijie, 9, 144–145.

*Lackner, M., Amelung, I., & Kurtz, J.* (2001). New terms for new ideas: western knowledge and lexical change in late imperial China (Vol. 52). Leiden: Brill Academic Publishers.

*Levenson, J. R.* (1964). Confucian China and its modern fate (Vol. 1). Berkeley: University of California Press.

*Levy, M. J.* (1966). Modernization and the structure of societies. Milan: Hoplie Editore.

*Lewis, N., & Weaver, A. J.* (2016). Emotional responses to social comparisons in reality television programming. Journal of media psychology, 28(2), 65–77.

*Li, G.* (2014). The future of the red cross on the background of trust crisis. Modern marketing, 2, 10–14.

*Li, S., Li, P., & Xiao, L.* (2006). Shichang xuqiu yu jihui xianjing-guanyu woguo xiaolingtong fazhan de sikao [Market demand and opportunity trap - Reflection on the development of Xiaolingtong]. Dianzi keji daxue xuebao (shehui kexueban), 8(3), 1–4.

*Li, Y.* (2008). Lun chengxiang eryuan tizhi gaige [A discussion about reform of urban and village]. Beijing daxue xuebao (zhexue shehui kexue ban), 2(6), 19-23

*Li, Y, & Wang, H.* (2007). Lun gaige huji zhidu dui nongmin shehui baozhang de yingxiang [Influence of hukou reform to welfare of rural villagers]. Gaige luntan, 7-10.

Liang, S. (2011). Zhongguo Wenhuayaoyi [Essential of Chinese culture]. Shanghai: Shanghai renmin chubanshe.

Lievrouw, L. A., & Farb, S. E. (2003). Information and equity. Annual review of information science and technology, 37(1), 499–540.

Lingard, B. (2010). Policy borrowing, policy learning: testing times in Australian schooling. Critical studies in education, 51(2), 129–147.

Liu, P., & Chen, D. (1988a). The propaganda of Marx, Engels, Lenin, Stalin and Mao Zedong. Chengdu: Sichuan shehui kexue chubanshe.

Liu, P., & Chen, D. (1988b). The propaganda of Marx, Engels, Lenin, Stalin and Mao Zedong. Chengdu: Sichuan shehui kexue chubanshe.

Lu, X. (2004). Rhetoric of the Chinese cultural revolution: the impact on Chinese thought, culture, and communication. Columbia: University of South Carolina Press.

Luhmann, N., & Cross, K. (2000). The reality of the mass media. Stanford: Stanford University Press Stanford.

Lull, J. (2013). China turned on: television, reform and resistance. London: Routledge.

Lundby, K. (2009). Mediatization: concept, changes, consequences. New York: Peter Lang.

Luo, Y. (2004). A study about cultural tourism in QianDongnan Dong area. Guizhou Ethnic Studies, 3, 23.

Machlup, F. (1962). The production and distribution of knowledge in the United States (Vol. 278). Princeton: Princeton University Press.

Madrazo, B., & van Kempen, R. (2012). Explaining divided cities in China. Retrieved March 8, 2015 from https://doi.org/10.1016/j.geoforum.2011.07.004.

Mannheim, K. (2006). Idéologie et utopie. Pairs: Les Editions de la MSH.

Mao, T. T. (1934). Guanxin qunzhong shenghuo, zhuyi gongzuo fangfa [Care about the life of the people and use good strategy]. Beijing: Renmin chubanshe.

Mao, T. T. (1938). Lun chijiu zhan [Discussion about long-term fight]. Beijing: Renmin chubanshe.

Mao, T.T. (1945). China's new democracy. New York: New Century Pub.

Mao, T. T. (1949). On the people's democratic dictatorship. Retrieved June 9, 2016, from https://www.marxists.org/reference/archive/mao/selected-works/volume-4/mswv4_65.htm.

Mao, T. T. (1983). Selections from Mao Tse-tung's opinion on the press. Beijing: Xinhua chubanshe.

Mao, T. T. (1991). Selection from Mao Tse-tung (Vol. 4(35)). Beijing: Renmin chubanshe.

Maspero, H. (1981). Taoism and Chinese religion. Amherst: University of Massachusetts Press Amherst.

May, P. J. (1992). Policy learning and failure. Journal of public policy, 12(04), 331–354.

McCarthy, J. D., & Zald, M. N. (1977). Resource mobilization and social movements: A partial theory. American journal of sociology, 82(06) 1212–1241.

McCombs, M. E., Shaw, D. L., & Weaver, D. H. (1997). Communication and democracy: exploring the intellectual frontiers in agenda-setting theory. London: Psychology Press.

McLeod, J. M., Daily, K., Guo, Z., Eveland, W. P., Bayer, J., Yang, S., & Wang, H. (1996). Community integration, local media use, and democratic processes. Communication research, 23(2), 179–209.

McQuail, D. (1994). Mass communication. Hoboken: Wiley Online Library.

McQuail, D. (2000). Some reflections on the western bias of media theory. Asian journal of communication, 10(2), 1–13.

McQuail, D. (2010). McQuail's mass communication theory. New York: Sage publications.

Mead, G. H. (1934). Mind, self and society (Vol. 111). Chicago: Chicago University of Chicago Press.

Meng, X., Zhang, R., & Shen, S. (2004). Chengshi fazhan jinchengzhong de ni chengshihua qushi jiqi jingjixue fenxi [The trend of anti-urbanization in the urbanization process with economical approach]. Jingji jingwei, 1(64), E171.

Meseguer, C. (2005). Policy learning, policy diffusion, and the making of a new order. The annals of the American academy of political and social science, 598(1), 67–82.

MIIT. (2016). Database of Ministry of Industry and Information Technology of People's Republic of China. Retrieved November 26, 2015, from http://www.gov.cn/shuju/index.htm.

Mittler, B. (2004). A newspaper for China?: power, identity, and change in Shanghai's news media, 1872–1912. Cambridge: Harvard University Asia Center.

Molnar, P. (2015). Free speech and censorship around the globe. Budapest: Central European University Press.

Mote, F. W., & Rogers, D. (1988). Intellectual foundations of China. New York: McGraw-Hill.

*Mu, Q.*, & *Lee, K.* (2005). Knowledge diffusion, market segmentation and technological catch-up: The case of the telecommunication industry in China. Research policy, 34(6), 759–783.

*Munro, D. J.* (1969). The concept of man in early China. Stanford: Stanford University Press.

*Mytelka, L. K.*, & *Smith, K.* (2002). Policy learning and innovation theory: an interactive and co-evolving process. Research policy, 31(8), 1467–1479.

*Navarro, Z.* (2006). In search of a cultural interpretation of power: the contribution of Pierre Bourdieu. IDS bulletin, 37(6), 11–22.

*NBSC.* (2016). Data from National Bureau of Statistics of China. Retrieved May 10, 2016, from http://www.stats.gov.cn/tjsj/.

*Neiger, B. L., Thackeray, R., Van Wagenen, S. A., Hanson, C. L., West, J. H., Barnes, M. D.,* & *Fagen, M. C.* (2012). Use of social media in health promotion purposes, key performance indicators, and evaluation metrics. Health promotion practice, 13(2), 159–164.

*Netto, A.* (2002). Media freedom in Malaysia: The challenge facing civil society. Media Asia-Singapore, 29(1), 17–23.

Newspaper. (2009). "Baozhi de fazhan licheng ji qushi [The development and trends of newspaper]. Retrieved June 8, 2015, from http://www.docin.com/p-16997220.html.

*Newton, K.* (1999). Mass media effects: mobilization or media malaise? British journal of political science, 29(04), 577–599.

*Nordenstreng, K.* (1997). Beyond the four theories of the press. Media and politics in transition. Leuven: Acco.

*Norris, P.* (2001). Digital divide: civic engagement, information poverty, and the internet worldwide. Cambridge. Cambridge University Press.

*O'Shaughnessy, M.*, & *Stadler, J.* (2012). Media and society. Oxford: Oxford University Press.

*Olick, J. K.*, & *Robbins, J.* (1998). Social memory studies: From "collective memory" to the historical sociology of mnemonic practices. Annual review of sociology, 24: 105–140.

*Ostini, J.*, & *Ostini, A. Y.* (2002). Beyond the four theories of the press: a new model of national media systems. Mass communication and society, 5(1), 41–56.

*Ostrowski, S. w.* (1991). Ethnic tourism—focus on Poland. Tourism management, 12(2), 125–130.

*Park, M.-J.*, & *Curran, J.* (2000). De-westernizing media studies. London: Psychology Press.

*Qian, Y.* (2012). Tan "Guo Meimei meijian" yinfa de "cishan weiji [Discussion of crisis of charity aroused by Guo Meimei]. Xinwen shijie, 1, 88–89.

*Ragnedda, M., & Muschert, G. W.* (2013). The digital divide: the internet and social inequality in international perspective (Vol. 73). London: Routledge.

*Rao, S. S.* (2005). Bridging digital divide: efforts in India. Telematics and informatics, 22(4), 361–375.

*Reiser, R., & Weiss, K.* (1969). Production of staphylococcal enterotoxins A, B, and C in various media. Applied microbiology, 18(6), 1041–1043.

*Rice, N. M.* (2015). Russian anti-Americanism, public opinion and the impact of the state-controlled mass media. Retrieved August 18, 2016, from http://trace.tennessee.edu/cgi/viewcontent.cgi?article=5058&context=utk_graddiss.

*Robison, K. K., & Crenshaw, E. M.* (2002). Post-industrial transformations and cyber-space: a cross-national analysis of Internet development. Social Science Research, 31(3), 334–363.

*Rogers, E. M.* (2010). Diffusion of innovations. New York: Simon and Schuster.

*Rose, R.* (1991). What is lesson-drawing? Journal of public policy, 11(1), 3–30.

*Sabatier, P. A.* (1986). Top-down and bottom-up approaches to implementation research: a critical analysis and suggested synthesis. Journal of public policy, 6(1), 21–48.

*Sahay, S.* (2016). Are we building a better world with ICTs? Empirically examining this question in the domain of public health in India. Information technology for development, 22(1), 168–176.

*Salomon, R., & Jin, B.* (2008). Does knowledge spill to leaders or laggards? Exploring industry heterogeneity in learning by exporting. Journal of international business studies, 39(1), 132–150.

*Schein, L.* (2000). Minority rules: the Miao and the feminine in China's cultural politics. Durham: Duke University Press.

*Schiffrin, H. Z.* (1968). Sun Yat-sen and the origins of the Chinese revolution. Berkeley: University Of California Press.

*Schradie, J.* (2011). The digital production gap: the digital divide and Web 2.0 collide. Poetics, 39(2), 145–168.

*Schramm, W.* (1964). Mass media and national development: the role of information in the developing countries. Redwood City: Stanford University Press.

*Schroeter, R.* (2012). Engaging new digital locals with interactive urban screens to collaboratively improve the city. Paper presented at the Proceedings of the ACM 2012 conference on Computer Supported Cooperative Work. 2012 February 11 – 15; New York, USA

Schultz, J. (1998). Reviving the fourth estate: democracy, accountability and the media. Cambridge: Cambridge University Press.

Schwarcz, V. (1986). The Chinese enlightenment: intellectuals and the legacy of the May Fourth Movement of 1919. Berkeley: University of California Press.

Shen, B. (1997). Zhenli biaozhun wenti taolun shibo [(The background of discussion about the 'truth']. Beijing: Zhongguo qingnian chubanshe.

Shen, J. (2005). Dianxin chongzu bingfei dianxin gaigede biran xuanze [Telecom reform is not an inevitable choice of China's telecommunication reform]. Zhongguo xin tongxin, 7, 5–6.

Siebert, F. S., & Schramm, W. (1956). Four theories of the press: the authoritarian, libertarian, social responsibility, and Soviet communist concepts of what the press should be and do. Illinois: University of Illinois Press.

Silverstone, R., & Haddon, L. (1996). Design and the domestication of ICTs: technical change and everyday life. Communicating by design: the politics of information and communication technologies, 44–74. Oxford: Oxford University Press.

Simmel, G. (1908). Soziologie: untersuchungen über die formen der vergesellschaftung. Berlin: Duncker & Humblot.

Simmons, B. A., & Elkins, Z. (2004) The globalization of liberalization: policy diffusion in the international political economy. American political science review, 98(01), 171–189.

Sina. (2014). "Market share of telecommunications companies after reuniting in 2000". Retrieved May 10, 2014, from http://tech.sina./com.cn/it/telecom/2000-09-26/37721.shtml.

Sorokin, P. A. (1941). The crisis of our age the social and cultural outlook. London: Dutton.

Stone, D. (1999). Learning lessons and transferring policy across time, space and disciplines. Politics, 19(1), 51–59.

Stone, D. (2001). Learning lessons: policy transfer and the international diffusion of policy ideas. Retrieved May 18, 2015, from http://wrap.warwick.ac.uk/2056/1/WRAP_Stone_wp6901.pdf.

Sun, L. (2004). Zhuanxing yu duanlie: gaige knifing yilai zhongguo shehui jingo de bianqian [Transform and fragmented society: the changing society after the reform and opening]. Beijing: Tsinghua University Press.

Sun, Y. (2002). Newspaper in China. Beijing: China Sancta Press.

*Sun, Y.* (2012). Lost from freedom to the cross-road----carry on cultural analysis for the historic and the mental state of Chinese modern liberalism intellectuals. Jinan: Shandong University Press.

*Swanson, L. A.* (1996). People's advertising in China: A longitudinal content analysis of the People's Daily since 1949. International journal of advertising, 15(3), 222–238.

*Talja, S.* (2005). The social and discursive construction of computing skills. Journal of the American society for information science and technology, 56(1), 13–22.

*Tan, S.-H.* (2008). Modernizing confucianism and 'new confucianism'. In K. Louie (Ed.), The Cambridge companion to modern Chinese culture, 135–154.

*Telegeography.* (2016). "Report of global internet geography". Retrieved February 6, 2017, from https://www.telegeography.com/page_attachments/products/website/research-services/global-internet-geography/0006/8538/Japan_Country_Profile.pdf.

*Television.* (2012). Zhongguo dianshiji fazhanshi [The history of television development]. Retrieved August 5, 2014, from http://www.doc88.com/p-035415184773.html.

*Television.* (2016). Qianxi dianshi jishu fazhan licheng [The development of television technology]. Retrieved June 3, 2014, from http://wenku.baidu.com/view/8daac980e45c3b3566ec8b8a.html.

*Tipps, D. C.* (1973). Modernization theory and the comparative study of national societies: a critical perspective. Comparative studies in society and history, 15(02), 199–226.

*Toffler, A., Longul, W., & Forbes, H.* (1981). The third wave. New York: Bantam books New York.

*Tu, W.* (1999). Confucius: the embodiment of faith in humanity. The world & I, 14(11), 292.

*Tu, W.* (2001). The ecological turn in new Confucian humanism: implications for China and the world. Daedalus, 130(4), 243–264.

*TV.* (2012). Zhongguo dianshi fazhan gaikuang [The description of television development]. Retrieved October 6, 2014, from (http://www.docin.com/p-492872312.html.

*Uri, P. M.* (1977). The information economy: definition and measurement. Retrived May 20, 2015, from https://files.eric.ed.gov/fulltext/ED142205.pdf.

*Van den Berghe, P. L.* (1994). The quest for the other: ethnic tourism in San Cristóbal, Mexico. Seattle: University of Washington Press.

van Deursen, A., & van Dijk, J. (2010). Internet skills and the digital divide. New media & society, 13(6), 893-911.

Van Deursen, A., & Van Dijk, J. (2011). Internet skills and the digital divide. New media & society, 13(6), 893-911.

Van Deursen, A. J., van Dijk, J. A., & Peters, O. (2011). Rethinking Internet skills: the contribution of gender, age, education, Internet experience, and hours online to medium-and content-related Internet skills. Poetics, 39(2), 125-144.

Van Dijk, J., & Hacker, K. (2003). The digital divide as a complex and dynamic phenomenon. The information society, 19(4), 315-326.

Van Dijk, J. A. (2005). The deepening divide: inequality in the information society. New York: Sage Publications.

Van Dijk, J. A. (2006). Digital divide research, achievements and shortcomings. Poetics, 34(4), 221-235.

Van Dijk, J. A. (2013). A Theory of the Digital Divide. In *Ragnedda, M. & Glenn W. Muschert, G.W* (Eds.), The Digital Divide (pp: 29-51). London: Routledge.

Vehovar, V., Sicherl, P., Hüsing, T., & Dolnicar, V. (2006). Methodological challenges of digital divide measurements. The information society, 22(5), 279-290.

Volland, N. (2003). The Control of the Media in the People's Republic of China. Retrieved November 10, 2016, from http://archiv.ub.uni-heidelberg.de/volltextserver/8048/

Wagner, R. G. (2008). Joining the global public: word, image, and city in early Chinese newspapers, 1870-1910. Albany: SUNY Press.

Wang, X. (2008). Xiaolingtong fazhan zhilu de lixing sikao [Rational thinking about development path of Xiaolingtong]. Tongxin shijie, 22, 15-16.

Wang, X. (2009). Mou Zongsan yu Yin Haiguang----Jianlun dangdai xinrujia yu ziyouzhuyi (Xia) [Mou Zongsan and Yin Haiguang----Discussion between new confucianism and liberty: the second part]. Tongren xueyuan xuebao, 4, 1-11.

Wang, Y. (2001). Zhongguo chengshihua daolude xuanze he zhang'ai [Choices and obstacles of China's urbanization]. Zhanlv yu guanli, 1, 31-37.

Wanifra. (2010), (2011), (2012). "Report from International Federation of Audit Bureaux of Circulations and World Association of Newspapers and News Publishers from 2010 to 2012". Retrieved May 9, 2014, from http://www.wan-ifra.org/microsites/world-press-trends.

Watson, J., & Hill, A. (2015). Dictionary of media and communication studies. New York: Bloomsbury Publishing.

*Webster, F.* (2006). The information society revisited. Handbook of New Media: Student Edition, 443.

*Wedam, W. F.* (1989). Dianshide fazhan qushi [The trend of television development]. Dianshi jishu, 8, 001.

*Wei, L., & Hindman, D. B.* (2011). Does the digital divide matter more? Comparing the effects of new media and old media use on the education-based knowledge gap. Mass Communication and Society, 14(2), 216–235.

*Wei, L., & Zhang, M.* (2006). The third digital divide: the knowledge gap on the internet. Journalism & communication, 4, 005.

*Wei, R.* (2008). Motivations for using the mobile phone for mass communications and entertainment. Telematics and informatics, 25(1), 36–46.

*Willcock, H.* (1995). Japanese modernization and the emergence of new fictwn in early twentieth century China: a study of Liang Qichao. Modern Asian studies, 29(4), 817–840.

*Wimmer, R., & Dominick, J.* (2013). Mass media research. Boston: Cengage learning.

*Wong, K. S.* (1992). Liang Qichao and the Chinese of America: a re-evaluation of his "selected memoir of travels in the new world". Journal of American ethnic history, 11(4), 3–24.

*Wu, G.* (2006). Words, concepts and ideas showing freedom in Chinese traditional ideology. Jishou daxue xuebao (shehui kexue), 27 (1), 11-14.

*Wu, S.* (1997). Zhongguo dianshiju fazhan shigang [The history and structure of television drama in China]. Beijing: Beijing guangbo xueyuan chubanshe).

*Wu, Z.* (2014). Da Zainan [Big disaster]. Beijing: Nanwen Boya.

*Xia, Q., & Ye, X.* (2003). Cong shiyu dao xuanhua: 2003 nian 2yue dao 5yue guonei meiti "SARS weiji" baodao genzu [From silence to shouting: reports from Chinese media about SARS from January 2003 to May 2003]. Xinwen yu chuanbo, 10(2), 56–65.

*Xia, S.* (2006). The change of the masses mobilization in China. Weishi, 2, 16-19.

*Xiang, W.* (2009). Yidong hulianwang fazhande huigu yu zhanwang [The review and estimation of mobile internet]. Dianxin jishu, 1, 66–69.

*Xu, G.* (2004). The history of broadcasting and television in China. The broadcasting and television press, 5(1)104–105.

*Xu, H.* (2005). A study of the media of Guangzhou during the initial stage of the SARS outbreak. Guangzhou daxue xuebao, 4, 6.

Xu, J. (2003). The ten discussions about intellectual in China. Shanghai: Fudan University Press.

Xu, Y. (2002). Zhongguo gudai baozhi fazhan lunyao [History of ancient Chinese newspapers]. Hunan dahong chuanmei zhiye jishu xueyuan xuebao, 2(4), 33–38.

Yang, G. (1986). Zhongguo jindai baokan fazhan zhuangkuang [The development of press in early China]. Beijing: Xinhua chubanshe.

Yang, Z. (1997). Fuxing yu fazhan: qiandongnan miaozu shequ de bianqian taishi [Get prosperous and development: the change of Qiandongnan Miao community]. Xinan minzu daxue xuebao: renwen shehui kexueban, 4, 20–25.

Yao, Q. (2001). Zhongguo jindai baokanyede fazhan yu bainian shehui bianqian [The development and changes of newspaper in early China]. Shehui kanxue jikan, 6, 122–127.

Yao, X., & An, L. (2012). Functions of Weibo in public sphere----a case study of "Guo Meimei". Southeast, 8., 23-25

Ye, Y., & *Huang, R.* (2004). Zhongguo liudong renkou tezheng yu chengshihua yanjiu [The migration in China and urbanization policy]. Zhongguo renmin daxue xuebao, 2, 75–81.

Yin, J. (2003). Press freedom in Asia: new paradigm needed in building theories. Paper presented at the annual convention of the Association for Education in Journalism and Mass Communication, 2003 July 30-August 2; Kansas.

Yin, J. (2008). Beyond the four theories of the press: a new model for the Asian & the world press. Journalism & communication monographs, 10(1), 3–62.

Yin, S. (2011). Gudai Dubai yang [The research of Dibao]. Qufu: Qufu daxue chubanshe

Yu, K. (2010). Xinyimin yundong, gongmin shenfen yu zhidu bianqan: dui gaige kaifang yilai daguimo nongmingong jinchengde yizhong zhengzhixue jieshi [New migration movement, identity of citizenship and social system transformation: a political explantion of rural migration work in urban after the reform and opening]. Xinhua wenzhai, 10, 6–10.

Yung, K. K.-c. (2015). Cold war currents and Chinese emigré intellectuals, 1949–1960. Twentieth-century China, 40(2), 146–165.

Zhang, H., & *Zheng, W.* (2011). Dahua yidong tongxin [Research of telecommunication]. Beijing: Qinghua daxue chubanshe.

Zhang, L. (2013). Banzengnan yu quanli kongzhi [Difficulty to get certification and control of power]. Renmin gongan, 22, 017.

*Zhang, P.*, & *Liu, X.* (2010). Zhongguo jingji zhenzhang baogao (2009-2010)-chengshihua yu jingji zengzhang 2009-2010 [Report of economic development in China from 2009 to 2010]. Beijing: Shehui kexue chubanshe.

*Zhang, X.* (2006). Dui guhua tongxin fazhan qushide sikao [Reflections on the trends of fixed-phone]. Pubai keji, 1, 27–30.

*Zhao, L.* (2013). Meijie, huayu, quanli, shenfen "nongmingong' huayu kaogu yu shenfen shengchan yanjiu [Research of tradition about media, speak, authority, identity of farmer workers]. Hangzhou: Zhejiang daxue Chubanshe.

*Zhao, X.* (2012). Chongtu yu hezuo: weibo wenze zhongde duihua celue yanjiu [Conflict and cooperation: the role of Weibo in questioning and communicating]. Beijing: Zhongguo shehui kexueyuan yanjiushengyuan.

*Zhao, Y.* (2004). Zhongguo guangbo dianshi tongshi [The history of China's radio and television]. Beijing: Beijing guangbo xueyuan chubanshe.

*Zheng, C.* (2013). Jindai waiwen baozhi yanjiu zongshu [Research Of early Chinese foreign newspaper]. Xiandai jiaoji: xiabanyue, 12, 80–81.

*Zheng, H.*, & *Lu, Y.* (2002). Chengshizhong nongye hukou jiecengde diwei, zailiudong yu shehui zhenghe [Statues of rural villagers in city, social mobilization and social integration]. Jianghai Xuekan, 2, 88–93.

*Zheng, S.*, & *Zhang, X.* (2011). Jingji tizhi gaige yu zhongguo dianxin hangye zengzhang:1994-2007 [Economic reform and increase of telecommunication industry: 1994-2007]. Jingji yanjiu, 10, 67–80.

*Zhou, J.* (1945). The masses should talk more. Weekly democracy, 1, 3.

*Zhou, Q.* (2001). Shuwang jingzheng: zhongguo dianxinyede kaifang he gaige [Digital-net competition: Reform and opening of the telecommunications industry in China]. Beijing: Sanlian shudianshe.

*Zhu, Z.* (1994). Lun buman xianzhuang [Discuss about the unsatisfied situation]. Beijing: Zhongguo guangbo dianshi chubanshe.

www.ingramcontent.com/pod-product-compliance
Ingram Content Group UK Ltd.
Pitfield, Milton Keynes, MK11 3LW, UK
UKHW021834210426
5322IPUK00018B/260